CRTD-Vol. 20–1

AF703339

RISK-BASED INSPECTION – DEVELOPMENT OF GUIDELINES

VOLUME 1
GENERAL DOCUMENT

PREPARED BY

The Research Task Force on Risk-Based Inspection Guidelines

APPOINTED BY

The Codes and Standards Research Planning Committee of the ASME Center for Research and Technology Development

FOR

The ASME Council on Codes and Standards
The United States Nuclear Regulatory Commission
The National Board of Boiler and Pressure Vessel Inspectors
The Pressure Vessel Research Committee – Welding Research Council
American Nuclear Insurers
The Hartford Steam Boiler Inspection and Insurance Company
Industrial Risk Insurers
The American Petroleum Institute

REVIEWED AND EDITED BY

Steering Committee on Risk-Based Inspection Guidelines and an Independent Peer Review Committee

NRC GRANT NO. NRC-04-89-102

THE AMERICAN SOCIETY OF MECHANICAL ENGINEERS

☐ Two Park Avenue ☐ New York, N.Y. 10016

DISCLAIMER

This report was prepared as an account of work sponsored through the American Society of Mechanical Engineers (the Society) Center for Research and Technology Development by the United States Nuclear Regulatory Commission, the National Board of Boiler and Pressure Vessel Inspectors, the Pressure Vessel Research Committee – Welding Research Council, the American Nuclear Insurers, The Hartford Steam and Boiler Inspection and Insurance Company, the Industrial Risk Insurers, and the American Petroleum Institute (collectively referred to herein as the Sponsors). Neither the Society, nor the Sponsors, nor Westinghouse Electric Corporation, the University of Maryland, Battelle Pacific Northwest Laboratories, Failure Analysis Associates, Inc., Factory Mutual Research Corporation, and the McDonnell Aircraft Company (collectively referred to herein as the Sponsorees), nor any agency thereof, nor any of their respective employees makes any warranty, express or implied, or assumes any legal liability or responsibility for the accuracy, completeness, or usefulness of any information, apparatus, product, or process disclosed, or represents that its use would not infringe upon privately owned rights. Reference herein to any specific commercial product, process, or service by trade name, trademark, manufacturer, or otherwise does not necessarily constitute or imply its endorsement, recommendation, or favoring by the Society, the Sponsors, the Sponsorees, or any agency thereof. The views and opinions of the authors, contributors, and reviewers of the report expressed herein do not necessarily reflect those of the Society, the Sponsors, the Sponsorees, or any agency thereof.

Statement from By-Laws: The Society shall not be responsible for statements or opinions advanced in papers . . . or printed in its publications (7.1.3)

Reprinted with corrections 2015.

ISBN No. 0-7918-0618-9
ISBN 13 : 978-0-7918-0618-0

Library of Congress
Catalog Number 91-55452

Copyright © 1991 by
THE AMERICAN SOCIETY OF MECHANICAL ENGINEERS
All Rights Reserved
Printed in U.S.A.

FOREWORD

In 1985, the American Society of Mechanical Engineers formed a Risk Analysis Task Force, chaired by Dr. Alan Moghissi under the direction of the Council on Engineering, in response to a need to initiate the use of risk-based methods in the formulation of policies, codes, and standards. At the suggestion of that task force, the ASME Codes and Standards Research Planning Committee of the ASME Center for Research and Technology Development recommended that a research program be initiated to determine how risk-based methods could be used to establish inspection requirements and guidelines for systems and components of interest to the engineering community. The committee initiated the project out of recognition that recent catastrophic structural failures, occurring across many industries, highlight the need for society to relate risk more explicitly with inspection programs, particularly as most industries are forced into life-extension practices because of current economic conditions.

The task force of recognized experts with the requisite background and experience from a broad range of industries and applications was formed during 1986 and 1987. This group provided a means for a cooperative, multi-disciplinary research effort to be performed that would have government, university, and industry participation. The task force also provided an independent framework for developing and recommending appropriate methods to other research programs and to codes and standards organizations of ASME and other engineering organizations.

Direct financial sponsorship began with a U.S. Nuclear Regulatory Commission research grant in late 1988 and was followed by seed money from the ASME Council on Codes and Standards and the National Board of Boiler and Pressure Vessel Inspectors in 1989. Indirect financial sponsorship has been provided by the respective organizations of each task force member since their efforts are only partially supported. Additional direct sponsorship was obtained in early 1990 from the Pressure Vessel Research Committee – Welding Research Council, the American Nuclear Insurers, Industrial Risk Insurers, The Hartford Steam Boiler Inspection and Insurance Company, and the American Petroleum Institute to support the long-term needs of the program.

The research task force has been meeting four times annually since late 1988. The group has sought knowledge from several sources: ongoing research within the respective organizations of each task force member (e.g., the inspection reliability program for nuclear power components at the Battelle Pacific Northwest Laboratory); information from other domestic and foreign research programs (e.g., an advanced risk-based inspection strategy, championed by Vic Chapman of Rolls-Royce and Associates Limited, that has been implemented in the United Kingdom Nuclear Submarine program); experience from working with code committees that are developing new inspection requirements (e.g., ASME Section XI Subgroup on Core Supports and Internals Structures); and related technical literature. The applications that have been explored range from inspection of power plant components to aircraft and marine ship hull structures.

This general document, Volume 1, describes and recommends appropriate processes and methods using risk-based information to establish inspection guidelines for facilities or structural systems. This general document is to be used in conjunction with supplemental volumes that apply to specific types of systems and which are currently under preparation for several applications. All of these documents may be employed by users in the development and implementation of their inspection programs. These guidelines may be used by code groups to prepare new or revised codes and standards. However, further results from pilot applications may be required to provide the technical basis for actual codes and standards changes.

It is hoped by everyone involved with the research project that code committees and other engineering organizations find this work to be useful in the development of inspection guidelines or requirements and in their own research efforts. Additionally, we hope that the suggested processes enhance communication among the many disciplines that are involved in developing and performing inspection programs for structural systems.

Kenneth R. Balkey
Chairman, ASME Research Task Force on
Risk-Based Inspection Guidelines

ACKNOWLEDGMENTS

Although this document represents the work of the research task force members, this study would not be possible without the contributions of a large number of leaders in their respective fields from academia, government, and industry.

The steering committee members have carefully guided the project, and the independent peer review members teamed with the steering committee to diligently review and edit this document. The valuable and generous contribution of these members, who are identified in this document, is most appreciated.

The research task force acknowledges with appreciation the contributions of Vic Chapman of Rolls-Royce and Associates Limited, who attended most of the meetings and provided input to the recommended processes in the document related to statistical approaches for updating the state-of-knowledge following inspection of components. Ahmed Ibrahim, a Ph.D. candidate at the University of Maryland, is acknowledged for his contributions regarding the treatment of uncertainties and in providing thorough examples of how to rank components based on quantitative risk measures. Comments by Dr. Lee Abramson of the U.S. Nuclear Regulatory Commission regarding the elicitation of expert opinion and risk-based ranking methodologies were much appreciated. Dr. Robert Perdue of the Westinghouse Electric Corporation's Science and Technology Center is cited for his valuable assistance in preparing the decision analysis example for choosing an inspection strategy.

The research task force also acknowledges with appreciation the presentations made at task force meetings by the following persons:

- Sture O. Angelsen, DNV Industrial Services, Inc., Houston, Texas
- Bruce A. Bishop, Westinghouse Electric Corporation, Pittsburgh, Pennsylvania
- Garry S. Holman, Lawrence Livermore National Laboratory, Livermore, California
- Len Katz, Consultant, ASME Section XI Working Group on Plant Life Extension, Pittsburgh, Pennsylvania
- Truong V. Vo, Battelle Pacific Northwest Laboratories, Richland, Washington
- Dr. Walter O. Wunderlich, Consulting Engineer, American Society of Civil Engineers, Knoxville, Tennessee

Finally, the Westinghouse Energy Center Word Processing staff and the ASME Technical Publishing Department members are acknowledged for their dedicated and diligent efforts in compiling, editing, and publishing this document.

EXECUTIVE SUMMARY

Based on a multi-disciplinary research effort, risk-based processes and methods for developing inspection guidelines for facilities or structural systems are recommended and described.

A four-part, risk-based inspection process is recommended to rank or classify systems, components, or elements for inspection and to develop the strategy (i.e., the frequency, method, and sample sizes) to perform the inspection. This process includes:

(a) definition of the system;

(b) a qualitative risk assessment;

(c) a quantitative risk analysis that includes an enhanced failure modes, effects, and criticality analysis (FMECA);

(d) the development of the inspection program for components and structural elements using decision risk-analysis methods.

This multi-disciplinary, top-down approach starts at the system level before focusing the inspection program at the component or structural element level. According to the recommended strategy, the results of the inspection should be continuously used to update the state-of-knowledge throughout the four parts of the approach so that a living process is achieved.

A key step in defining a system for inspection is the assembly of information that is needed for the risk-based approach. In particular, the interviewing of key personnel, who are knowledgable of degradation mechanisms or errors that may not be documented is vital to the process.

The qualitative risk assessment captures fundamental expert judgment and experience in prioritizing systems, components, or elements for inspection. A key element of this assessment is to thoroughly define failure modes and causes, including design, operational, and maintenance errors and a host of potential degradation mechanisms.

The FMECA captures the information from the qualitative risk assessment and assigns probabilities of failure and consequences for each component or element that is eligible for inspection. Once again, the elicitation of expert opinion from knowledgable personnel is a valuable source of information. The probability of failure for each element is multiplied by its respective consequence in order to obtain risk. This process can be performed using single point estimates, or uncertainties in the probability and consequence values can be treated with interval estimates by defining lower and upper estimates for these values. Various plots are suggested to perform the risk-based ranking of components or elements, which can then be grouped or categorized to facilitate the establishment of the inspection program.

A three-step process is recommended to develop the inspection program for each group or category of components or elements. The process consists of:

(a) choosing candidate inspection strategies;

(b) selecting an inspection strategy and performing the inspection;

(c) deciding on appropriate action based on the inspection results and then updating the state-of-knowledge.

Decision analysis logic is recommended in order to understand the approach. Structural reliability and risk assessment (SRRA) models are recommended to be exercised to determine the effect of inspection reliability, potential degradation mechanisms, and potential loadings on the probability of failure of the component. Inspections in themselves do not affect the actual failure probability of the component. Rather, inspection data provides a means of building up confidence in the level of safety of the component being inspected.

A methodology for determining this effect is offered. The SRRA models should also be exercised to evaluate corrective actions, relative to timing, potential for success, and potential for damage to be introduced, if significant findings occur. Regardless of what path is followed, these results are used to update the FMECA and the inspection strategy, and for some cases, redefinition of the system may be required.

A partial list of available software tools for performing parts of the risk-based inspection process is provided. However, it takes many tools to perform the entire process, and tools will have to be developed to meet this need, including those to advance the integration of cost-benefit analysis techniques with engineering risk-analysis methods for optimization of inspection programs. To support this development work, further research needs to be performed for applications of interest to the engineering community using the recommended risk-based inspection process. Further efforts are being performed first for nuclear power plant applications, which will be reported in Volume 2. Applications to other industries, such as fossil fuel-fired power plant and petroleum refinery processing and storage components, are also expected to occur in order that appropriate tools are developed to meet the needs of these industries.

The primary benefit of the risk-based inspection approach is the focusing of inspection efforts on systems and components associated with the highest risk. Additional significant benefits are the insights gained in working through these processes, and the enhancement in communications among the many disciplines that are involved to help maintain an adequate level of safety within the affected industries and society in general.

RESEARCH TASK FORCE ON RISK-BASED INSPECTION GUIDELINES

Kenneth R. Balkey, *Chairman,* P.E., Principal Engineer, Westinghouse Electric Corporation, Pittsburgh, Pennsylvania

Dr. Bilal M. Ayyub, P.E., Associate Professor, Department of Civil Engineering, University of Maryland, College Park, Maryland

Dr. Bryan F. Gore, Staff Scientist, Battelle Pacific Northwest Laboratories, Richland, Washington

Dr. David O. Harris, Managing Engineer, Failure Analysis Associates, Inc., Menlo Park, California

Dr. Dimitrios Karydas, P.E., Manager – Engineering Risk Analysis, Factory Mutual Research Corporation, Norwood, Massachusetts

Dr. Fredric A. Simonen, Technical Leader, Battelle Pacific Northwest Laboratories, Richland, Washington

Dr. Herb Smith, Jr., P.E., Technical Specialist – Structural Research, McDonnell Aircraft Company, St. Louis, Missouri

RESEARCH STEERING COMMITTEE ON RISK-BASED INSPECTION GUIDELINES

Raymond J. Art, Assistant Director, ASME Center for Research and Technology Development, Washington, D.C.

Robert J. Bosnak, ASME Council on Codes and Standards, ASME Codes and Standards Research Planning Committee, ASME Board on Research and Technology Development, Deputy Director, U.S. Nuclear Regulatory Commission, Washington, D.C.

Dr. Spencer J. Bush, Consultant, Past Chairman – ASME Section XI, Richland, Washington

Will J. Carter, Staff Engineer, Chevron Corporation, (American Petroleum Institute), Richmond, California

Theodore A. Meyer, Manager, Westinghouse Electric Corporation, Pittsburgh, Pennsylvania

Evangelos Michalopoulos, P.E., Senior Engineer, The Hartford Steam Boiler Inspection and Insurance Company, Hartford, Connecticut

Michael F. Moylan, Senior Project Engineer, Wisconsin Electric Power Company, Milwaukee, Wisconsin

Dr. Joseph Muscara, Senior Metallurgical Engineer, U.S. Nuclear Regulatory Commission, Washington, D.C.

Michael E. G. Schmidt, P.E., Research Consultant, Industrial Risk Insurers, Hartford, Connecticut

William G. Wendland, P.E., Manager – Engineering Projects, American Nuclear Insurers, Farmington, Connecticut

INDEPENDENT PEER REVIEW COMMITTEE

Steven W. Pullins, Pullins Engineering Company, Chillicothe, Ohio
Dr. C. (Raj) Sundararajan, EDA Consultants, Humble, Texas
Paul J. Torpey, Empire State Electric Energy Research Corporation New York, New York

CONTENTS

APPENDICES

INTRODUCTION 1

1.1 □ BACKGROUND

Inservice inspections (ISI) can play a significant role in preventing equipment and structural failures. All aspects of inspections, i.e., objectives, method, timing, and acceptance criteria, can affect the *likelihood* of component failure. However, for many pressure-boundary components or others that are subjected to various service conditions in the nuclear power, fossil fuel-fired power, and petroleum and chemical processing industries, as well as others, inservice inspection requirements are either based upon prior experience and engineering judgment or are nonexistent. Most inservice inspection requirements or guidelines are established with only an implicit consideration of risk-based information and analysis (i.e., the *probability* of failure for the plant-specific material, operation, and loading conditions, and of the *consequence* if component failure occurred).

Qualitative and quantitative approaches can be used for dealing with the concepts of *hazard* and *risk*. In the terminology of these approaches, risk is the measure of the potential for harm or loss (i.e., hazard) that reflects both the likelikhood (e.g., *frequency*) and severity of an adverse effect to health, property, or environment.

Much development work has focused on *probabilistic risk assessment (PRA),* which is an evolving technique for scientifically evaluating the probability and the impact of an adverse effect. This impact may be in the form of shock wave propagation, thermal loading, health effects, and/or environmental damage. The probability of an adverse effect is generally determined using "logic" trees (e.g., *fault tree analysis*) and branching decision networks (e.g., *event tree analysis*). PRA methodology has been primarily applied to the evaluation of *facility*-wide, or macroscopic, assessments as compared to the assessment of individual components of a system, or microscopic, assessment. The main benefit of PRAs has been to identify design or procedural changes to avoid safety hazards and prevent losses.

PRA technology, which has been used quite extensively in the nuclear industry following the U.S. Nuclear Regulatory Commission (NRC) Reactor Safety Study (WASH-1400) (1975), has been successfully applied in several other industries, as discussed by Moghissi, et al., in *Mechanical Engineering* (1984).

During the past ten years, *probabilistic structural mechanics (PSM)* is another technique that has emerged as a tool for assessing the structural risk and reliability of components and structures in many industries. One promising application of PSM methods is the establishment of nondestructive examination criteria. In contrast to PRA, PSM relates to the microscopic assessment of components to address the mechanistic uncertainties, such as those of stress, load, and material properties. [See Balkey, Meyer, and Witt in *Mechanical Engineering* (1986) for further discussion of PSM.] PRA and PSM methods are interrelated and may be used most advantageously in combination.

EDITORIAL NOTE: Definitions are provided in Section 3 of this document for terms or words shown in italics that appear for the first time in the text.

ASME has recognized the need for risk-based methods in the formulation of policies, codes, and standards. In 1985, under the direction of the Council on Engineering, a Risk Analysis Task Force, headed by Dr. Alan Moghissi, was formed by the Society to provide recommendations on how this need could be met. Recent catastrophic structural failures, such as pipe ruptures in fossil fuel-fired generating stations, tank failures in the processing industry, collapsing bridges, and the breakup of major aircraft components, highlight the need for the Society to relate risk more explicitly with inspection programs.

At the suggestion of the Risk Analysis Task Force, the ASME Codes and Standards Research Planning Committee (CSRPC) recommended in 1986 that a research program be initiated to determine how risk-based methods, such as PRA and PSM, could be used to develop inspection requirements and guidelines for systems and components of interest to the engineering community. During 1987, Mr. Kenneth Balkey, working with the ASME Center for Research and Technology Development and CSRPC, suggested a research program to meet this goal. The program was approved by the ASME Board on Research and Technology Development and the ASME Council on Codes and Standards in the fall of 1987. The Research Task Force members jointly developed a work plan in early 1988 that was put in place in late 1988 upon receipt of financial support from the project's first sponsor, the U.S. NRC.

Facility life extension is taking a greater importance in all industries due to the current economic climate. Since inservice inspections can play a significant role in preventing equipment and structural failures, they can also play a major role as part of residual life or life-extension assessments. For example, an ASME Code Section XI Working Group has been formed to define Code changes that may have to be made to accommodate nuclear power plant life extension using traditional *deterministic analysis* methods. However, research by the NRC and others has made extensive use of risk-based methods to address issues related to plant aging and life extension. Although both life extension evaluations and the research program go beyond traditional ASME codes and standards applications relative to construction materials and techniques, ASME is becoming increasingly involved in this work.

Finally, other engineering societies and organizations in other countries have work underway to apply risk-based methods in the development of codes and standards for the inspection of bridges, offshore platforms, and aircraft. The research program makes as much use as possible of this work.

1.2 □ OBJECTIVES AND SCOPE

The fundamental objective of this ASME research program is to determine appropriate risk-based methods for developing inspection guidelines. Furthermore, in certain areas, these methods will be used to define risk-based inspection programs for recommendation to ASME and other codes and standards bodies. These methods should be applicable to all areas where structural failures have the potential to result in loss or damage. These areas include systems and components in nuclear and fossil fuel-fired power stations, aircraft structures, civil engineering and marine structures, and many other industrial applications.

This research program is principally concerned with the structural integrity of systems and components or elements and not with operational requirements. Plants and facilities are designed and constructed so that, in general, risk is dominated by failures of active, operational components. Structural component inspections are performed to ensure that the risks from structural failures are maintained much lower than risks from operational component failures. However, many of the principles of the risk-based inspection processes could give insight to the analysis of operational considerations as well.

This general document, Volume 1, describes and recommends appropriate state-of-the-art processes and methods using risk-based information to develop inspection guidelines for facilities or structural systems. These methods assist in the identification of the need, scope, objectives, timing, and acceptance criteria for inservice inspection of systems and components designed and constructed to ASME or other industry codes and standards. Additional supplemental volumes will provide guidelines for the application of risk-based methods to specific areas of interest where structural integrity failures may pose significant risk.

Many of the concepts presented in this general document were first presented in a paper entitled, "Probabilistically-Based Inspection Guidelines," by Balkey, et al. (1986), which was presented at the Risk Analysis Forum as part of the 1986 ASME Winter Annual Meeting and served as a technical starting point for the project. The paper provided a summary review of current inspection requirements for systems and components in several industries and identified some of the risk-based methods that are currently available for the development of guidelines for cost-effective inspections.

In accordance with the research plan, Appendix A of this document presents a review that builds on the above-cited paper of current inspection requirements and related developments in the areas of nuclear power through industrial insurance applications. A comprehensive risk-based process has been developed for outlining the scope of inspections for components in any given system. This process, outlined in Section 2, uses an enhanced *failure modes, effects, and criticality analysis (FMECA)* methodology. The method for prioritizing components or system elements for inspection by combining the probability and consequences of component failure to calculate associated risk is discussed. A *decision analysis* process for defining the timing and approach that should be used to perform the inspection of an important component is also included in Section 2. Some needed definitions are given in Section 3. Some available software tools for developing a risk-based inspection program are summarized in Section 4. Recommendations for application of these methods and for further research work are given in Section 5. References to other reports and technical literature are given in Section 6, followed by additional appendices that provide more details on methodologies for further consideration.

This ASME research effort is thus using an interdisciplinary approach by integrating technologies from a broad base of applications to identify risk-based processes and methods for developing inspection guidelines that will benefit society in general. Use of risk-based inspection methods for applications including nuclear and fossil fuel-fired power plants, aircraft structures, civil engineering and marine structures, and industrial insurance applications typify some of the technology that is being explored in the research effort. Examples are provided throughout this general document.

1.3 ☐ IMPLEMENTATION

This general document, Volume 1, recommends appropriate risk-based process and methods for use in developing an inspection program for any facility or structural system. This general document is used in conjunction with supplemental volumes which apply to specific types of systems. For example, an ASME Volume 2 – Part 1 document (1991), which recommends specific methods for light water reactor (LWR) nuclear power plant components is currently being prepared.

Future efforts that are planned for this project include the development of additional volumes to recommend specific methods for fossil fuel-fired power plants (Volume 3); petroleum refining processing and storage facilities (Volume 4); noncommercial nuclear facilities (Volume 5) and other applications given the increasing interest for use of this technology. A Volume 2 – Part 2 document is also being planned to recommend a risk-based inspection program for LWR nuclear power plants, including the technical basis, for consideration by the appropriate groups of Section XI of the ASME Boiler and Pressure Vessel Code.

Figure 1-1 displays the organization and use of the reports for the research project. All of the documents that are prepared in this research effort may be employed by users in the development and implementation of their inspection programs. These documents may also be used by code groups to prepare new or revised codes and standards. However, further results from pilot applications may be required to provide the technical basis for actual codes and standards changes.

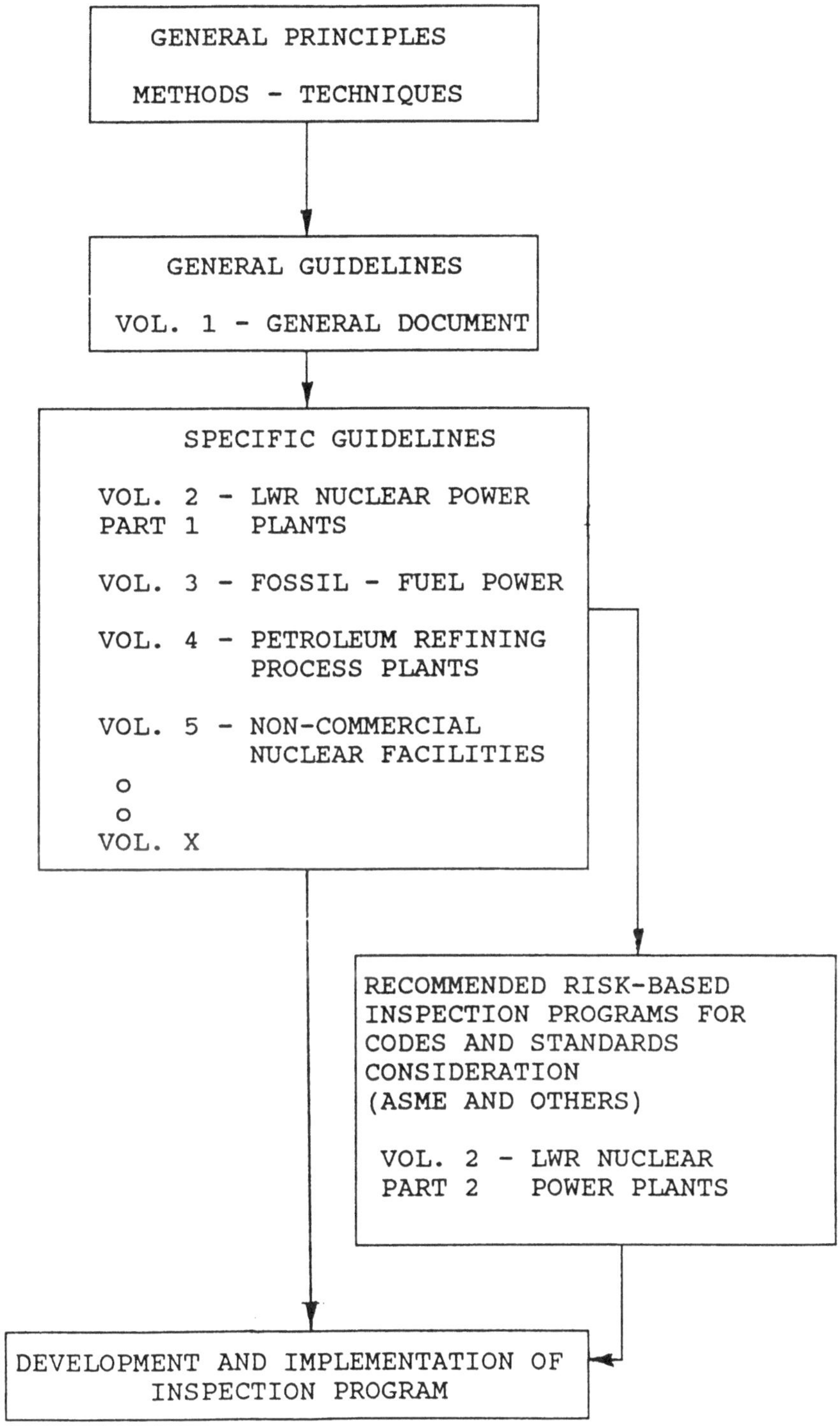

FIG. 1-1 OVERVIEW OF ASME RESEARCH PROJECT ON RISK-BASED INSPECTION GUIDELINES

OVERALL PROCESS FOR RISK-BASED INSPECTION

2

2.1 □ OVERVIEW

The development of a risk-based inspection process should include the prioritization of systems, subsystems, components, or elements for inspection using risk measures, and the definition of a strategy (i.e., the frequency, method, and sample sizes) for performing the actual inspections. The process should also include logic for making repair, replace, or do-nothing decisions following inspection. Finally, there should be a strategy for updating the inspection plan for a given structural system, subsystem, component, or element using the results of the inspections that are performed.

The important features of a risk-based inspection process should include:

- the use of a multi-disciplinary top-down approach that starts at the system level before focusing the inspection at the component or structural element level;
- the use of a "living" process that is flexible, strives for completeness, and can be easily implemented in order to control risk;
- the use of quantitative risk measures;
- the use of effective and efficient analytical methods that provide results that are readily reviewable and that are familiar to those involved in inservice inspection technology.

Figure 2-1 outlines the overall risk-based inspection process based on the needs defined above. The process is comprised of the following four parts:

- definition of the structural system that is being considered for inspection;
- use of a qualitative risk assessment that captures fundamental expert judgment and experience in identifying failure modes, causes, and consequences for initial ranking of systems and components for inspection;
- application of quantitative risk analysis methods, primarily using an enhanced FMECA and treating uncertainties, as necessary, to focus the inspection efforts on systems and components associated with the highest calculated risk;
- development of the inspection program for the components, using decision risk-analysis methods, beginning with an initial inspection strategy and ending with an update of that strategy based on the findings from the inspection that is performed.

Several feedback loops are shown to represent a living process for the definition of the system, the ranking of components or structural elements, and the inspection strategy for each component or element. A key objective is to develop a risk-based inspection process that is first established and then maintained up-to-date for the facility of interest.

The sections below provide information and examples for each part of the recommended process.

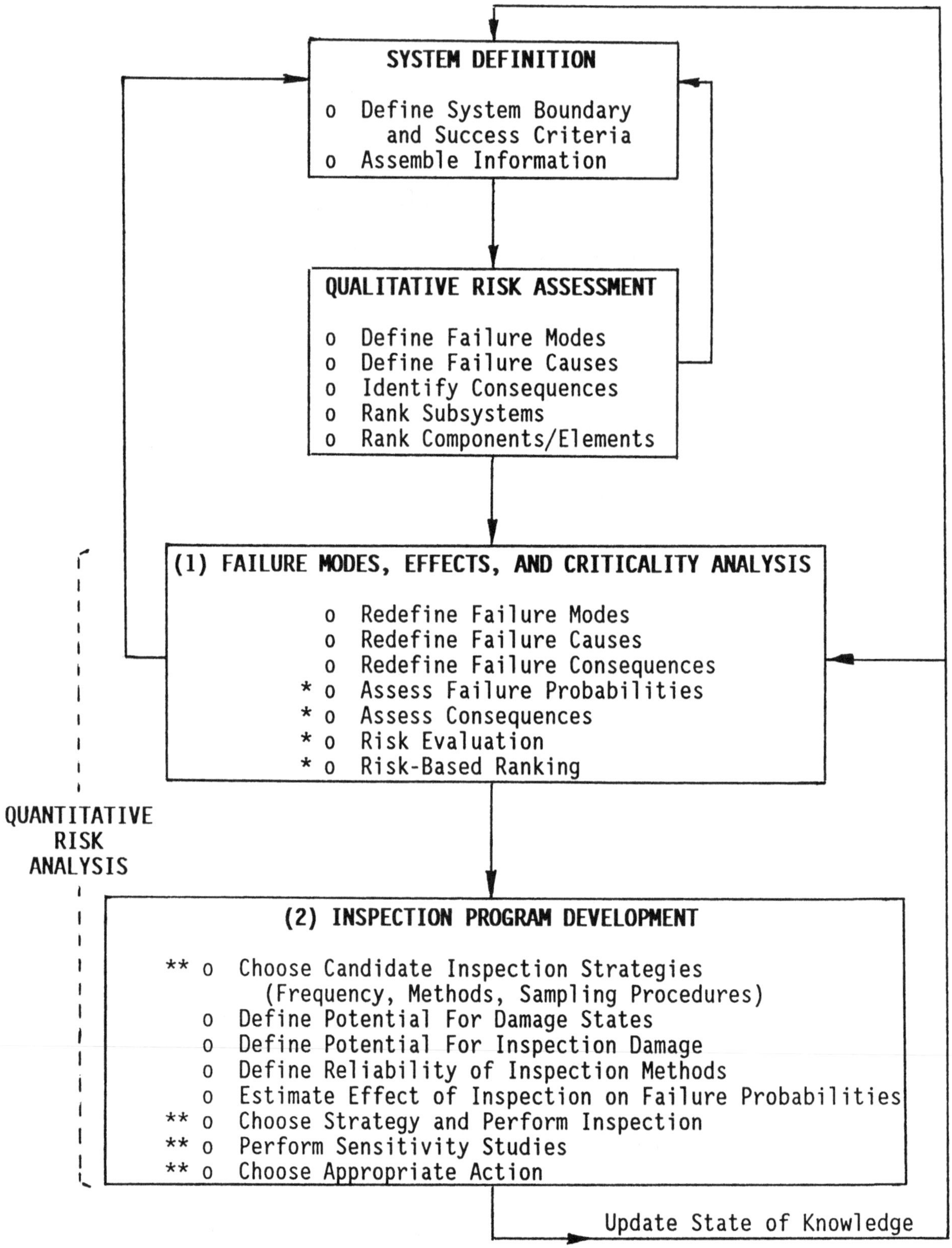

* With and Without Uncertainty
** Decision Risk Analysis

FIG. 2-1 RISK-BASED INSPECTION PROCESS

2.2 □ SYSTEM DEFINITION

2.2.1 □ Define System Boundary and Success Criteria

In order to develop an inspection strategy for any facility, the system must be defined and bounded, and its functions must be identified. Degradation must also be considered, and criteria must be specified that identify the minimum level of performance at which the system function is accomplished. The boundaries of the system are determined based on the nature or type of the facility, class of failures under consideration, and the objectives of the analysis.

In a general engineering sense, a physical system can be defined in the following two ways, according to Blanchard and Fabrycky (1981):

(1) an assemblage or combination of elements or parts forming a complex or unitary whole (e.g., a piping system); and

(2) an assemblage or set of correlated members (e.g., pressure vessels).

A system or subsystem can be described by its elements, which comprise the following:

- *Components* – the operating parts of a system which can be described by input, process, and output variables.
- *Attributes* – the properties of the components and/or characteristics of the system parameters.
- *Relationships* – the links between components and attributes.

The Institute of Electrical and Electronics Engineers (IEEE) (1984) defines electrical and mechanical systems in Nuclear Power Stations using all three of the above elements, i.e., on the basis of the System Function, System Description, Mode of Operation, Major Interfacing Systems, and Major Equipment included in the system.

Other entities characterize systems using only one or two of the above system description elements. For example, jurisdictional authorities are interested in pressure vessels, or firefighting systems, or environmental protection systems, etc. One object, or set of objects, may belong to one system or another depending on the definition and the scope of the system. Attributes of the same parameter may differ depending on the system definition. A nonpressure vessel, for example, containing hazardous material does not belong to the jurisdictionally defined system of pressure vessels and needs no inspection from that point of view. Environmental consequences may categorize this vessel to a different system with completely different attributes.

System definition is, therefore, the first step in identifying what components should be inspected and what the purpose should be of the inspection.

2.2.2 □ Assemble Information

A proper inspection program requires an adequate base of information. A major benefit from application of a detailed risk-based methodology is that the user is required to talk to the individuals who have first-hand knowledge and direct access to the information about the facilities, systems, and components to be inspected. Table 2-1 lists activities that are associated with this information-gathering effort. This effort represents a large part of the risk-based planning process, and its successful completion requires the participation and support of a wide range of people from the organization responsible for the facility of concern.

The first step is to perform an inventory of the relevant systems and components at the facility. This step has received considerable emphasis in recent discussions of inspection planning according to Clevenger (1989) and Dana, Sharp, and Webb (1989). For some facilities, such inventory lists may already exist, while in many cases, the list may need to be constructed. The objective is to systematically describe all the items of interest that may often be located at widely scattered sites within the overall boundaries of the facility.

As a rule, field surveys are required both to complete the inventory process and to verify data from available records. In some cases where records are limited, the field survey could be the primary source of information used to develop the inventory of systems and components.

The next step is to screen the inventory list for those items with structural implications, such as components that provide containment of high-energy fluid or hazardous material. This screen will identify the components of concern from the standpoint of this general document.

Once the inventory is completed, engineering records and documented operating procedures must be collected with the assistance of personnel at the facility. This includes engineering records that document the design, operating history, inspection, maintenance, repairs, and alterations. In some cases, results of prior risk studies (e.g., PRAs) may be available, and this information would also be most valuable in developing a risk-based inspection program.

Once a preliminary base of information has been collected and assimilated, in-depth interviews with cognizant personnel begin. These interviews serve to fill inevitable gaps in the engineering documentation. This step identifies potential failure modes and degradation mechanisms that could lead to structural integrity failures. Such information is used to perform failure modes, effects, and criticality analyses. The interview process also assists in making estimates of the consequences of structural integrity failures.

The final step indicated in Table 2-1 is to compile information that describes the experience at other sites with similar facilities. Structural integrity failures are generally rare events. Therefore, statistical estimates of failure probabilities require historical experience covering many years of operation for a large population of systems and components. This experience must additionally consider the potential for long-term aging effects. Potentially useful sources of information include both published literature on structural reliability, and statistical databases maintained by industry groups and regulatory agencies. For a given system or component, engineering judgment must be exercised in the application of generic information on failure probabilities, taking into account the operating history at the facility, age of the components, and the operating and maintenance experience for the component being addressed.

While the assembly of information represents one of the initial steps in the development of a risk-based inspection program, efforts to obtain information continue throughout the evaluation. Furthermore, once the inservice inspection is performed on a periodic basis, it is important to update the inservice inspection plan using new information. Such information comes from findings of the inspections as they are performed, accumulated service experience, maintenance and repair activities, and other sources such as industry databases and professional contacts.

2.3 □ QUALITATIVE RISK ASSESSMENT

2.3.1 □ Introduction

The next step in developing risk-based inspection guidelines for a given system or subsystem of components is to develop an inspection ranking model. Because of the costs associated with performing inspections, a need exists to focus inspection efforts on systems and components associated with the highest risk.

TABLE 2-1
STEPS FOR ASSEMBLY OF INFORMATION TO BE USED IN RISK-BASED INSPECTION

- List systems, subsystems, and components of potential concern.
- Obtain fluid system chemistry information, system environment, and system arrangement (isometric) to review system rigidity.
- Inventory components for the systems of interest at the facility.
- Identify those components for which structural integrity failures could have potential safety and/or economic consequences.
- Collect engineering records (design data, risk studies, operating history, inspection, maintenance, repairs, and alterations) and documented procedures.
- Conduct field surveys to verify engineering records, establish as-built configuration, and look for deviations/degradation from design conditions.
- Interview cognizant personnel to compile nonrecorded information and establish operational practices.
- Compile operational and failure experience at other sites and facilities (from literature, industry groups, professional contacts, regulatory agencies, etc.).

Most of the methods used for assessing risk of complex systems are based on a combination of analytical (i.e., quantitative) and judgmental (i.e., qualitative) approaches. For certain objectives and systems, however, a quantitative risk analysis can be too expensive, time consuming, or inconvenient. In such cases, a qualitative assessment can be more appropriate, especially where insufficient data is available to analytically assess the system. In addition, qualitative methods can be used to perform a pre-assessment of the system (i.e., before performing a detailed quantitative and qualitative evaluation of the system). Engineering judgment and experience are considered to be the bases of the qualitative analysis of consequences and likelihoods of failure. Therefore, the results are dependent on the background and expertise of the analysts and the objectives of the analysis.

2.3.2 □ Qualitative Analysis Methodology

The process of qualitative analysis is divided into five main categories:
(1) system identification and description;
(2) identification of initiating failure events;
(3) identification of consequences;
(4) qualitative estimation of risk levels considering both consequences and likelihoods of failure;
(5) ranking of subsystems and components for inspection purposes.

The first three categories are considered to be parts of a learning loop. For these categories, a qualitative flow chart can be constructed in order to model a system for the purpose of ranking its failure modes, components, and subsystems. The qualitative flow chart can be of two types, bottom-up or top-down. General flow charts representing the two types are shown in Figs. 2-2 and 2-3, respectively.

In the bottom-up approach, failures of individual components are initially identified, then the effects of these failures in certain sequences on the system, as a whole, are assessed. In the top-down approach, identified gross losses of system functions are decomposed to failure sequences of individual components of the system. Depending on the application, either approach or a combination of both may be appropriate.

Before performing the qualitative analysis, according to one of the two types of flow charts, the system should be defined as previously described in Section 2.2. In the case of the bottom-up type (i.e., Fig. 2-2), the first step in the analysis is to establish limits on consequences (e.g., in terms of casualties and economic loss). These limits depend on the objective of the analysis and any constraints imposed on the analyst. The limits provide upper and lower bounds on the consequences. Then, all possible initiating events that lead to defined failure modes should be identified. The definition of the initiating events and failure modes is expected to be heavily based on available performance records of the system, and the judgment of experts and personnel who are intimately knowledgable of the system. The consequences of failure should be qualitatively assessed and compared to the established limits. Based on this analysis, the system may require redefinition and refinement through a learning loop. The subsystems should then be defined along with any redundancies. Similarly, the components within each subsystem should be defined. Any redundancies within the subsystem based on the composition and interaction of its components should be identified. The final task is to perform the qualitative-based ranking of the components according to their failure modes and the qualitative-based ranking of the subsystems. An additional product of the process is a definition of the failure modes of the system, which is discussed later in this section in detail.

In the case of the top-down strategy (i.e., Fig. 2-3), the first step in the analysis is to identify all possible failure modes of the system under investigation. Then, subsystems and redundancy in the subsystem level should be identified. For each subsystem, the failure modes should be identified. Choosing a failure mode and defining the components of that subsystem is the next step. For each component, the failure modes should be defined. Selecting a component failure mode and defining its initiating events and consequences can be used as criteria for ranking the failure modes of the components. The susbsystems and the failure modes of the subsystems can be ranked according to the likelihood of failure, consequences, or risk when consequence estimates have been completed for all subsystem failure modes. Figure 2-4 offers a suggested presentation that has been adapted from approaches used by the U.S. Environmental Protection Agency (EPA) (1987) and by Lercari (1989) for the State of California Office of Emergency Services.

2.3.3 □ Define Failure Modes and Causes

Before an inspection strategy can be established, it is first necessary to identify the potential modes and causes of failure for the structural components of concern. Furthermore, one must relate failure modes to possible safety and/or economic consequences. At this stage of the evaluation, the foremost objective is to ensure that all significant failure modes and associated causes are addressed for later use in the risk-

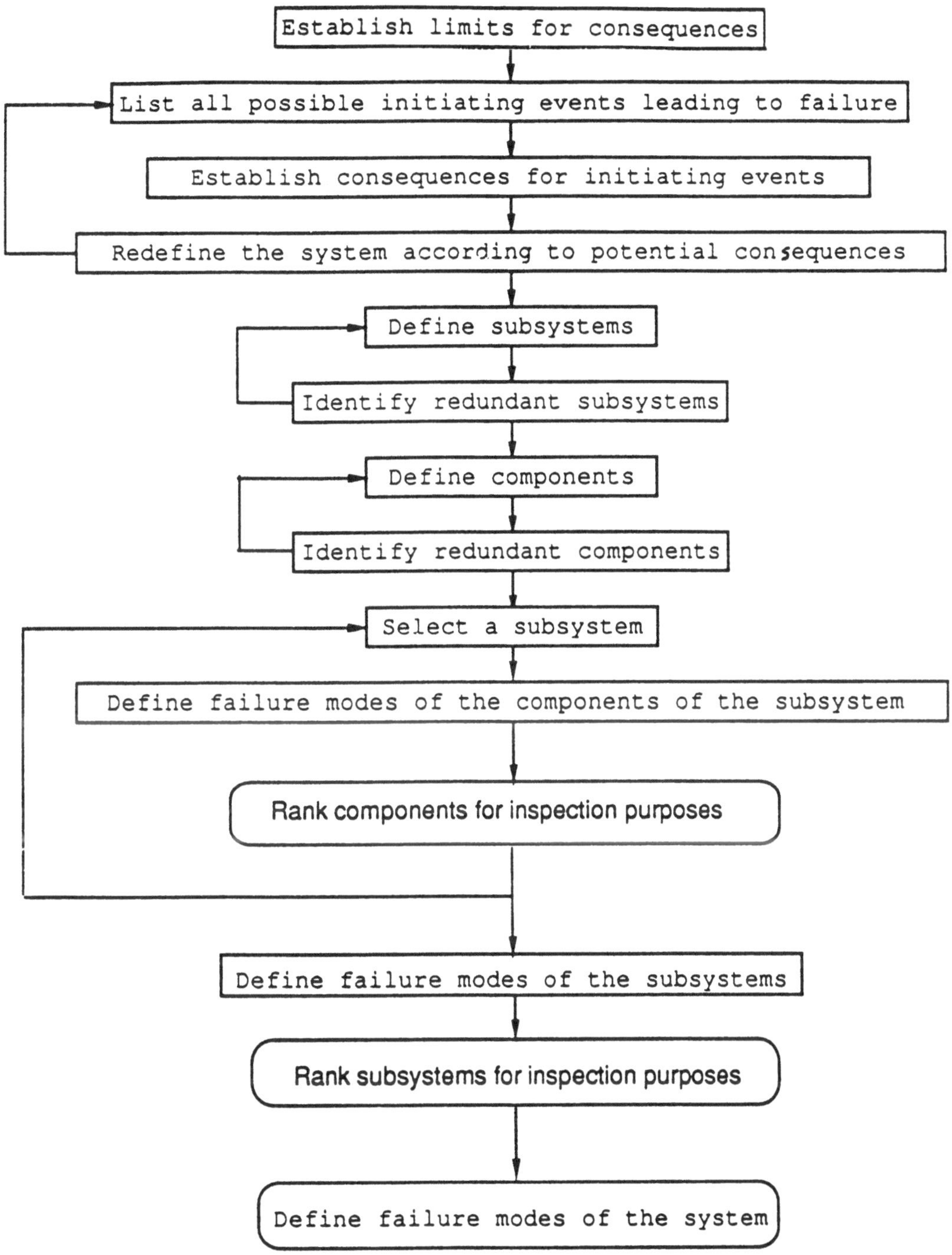

FIG. 2-2 BOTTOM-UP QUALITATIVE ANALYSIS FOR INSPECTION RANKING AND FAILURE MODE DEFINITION

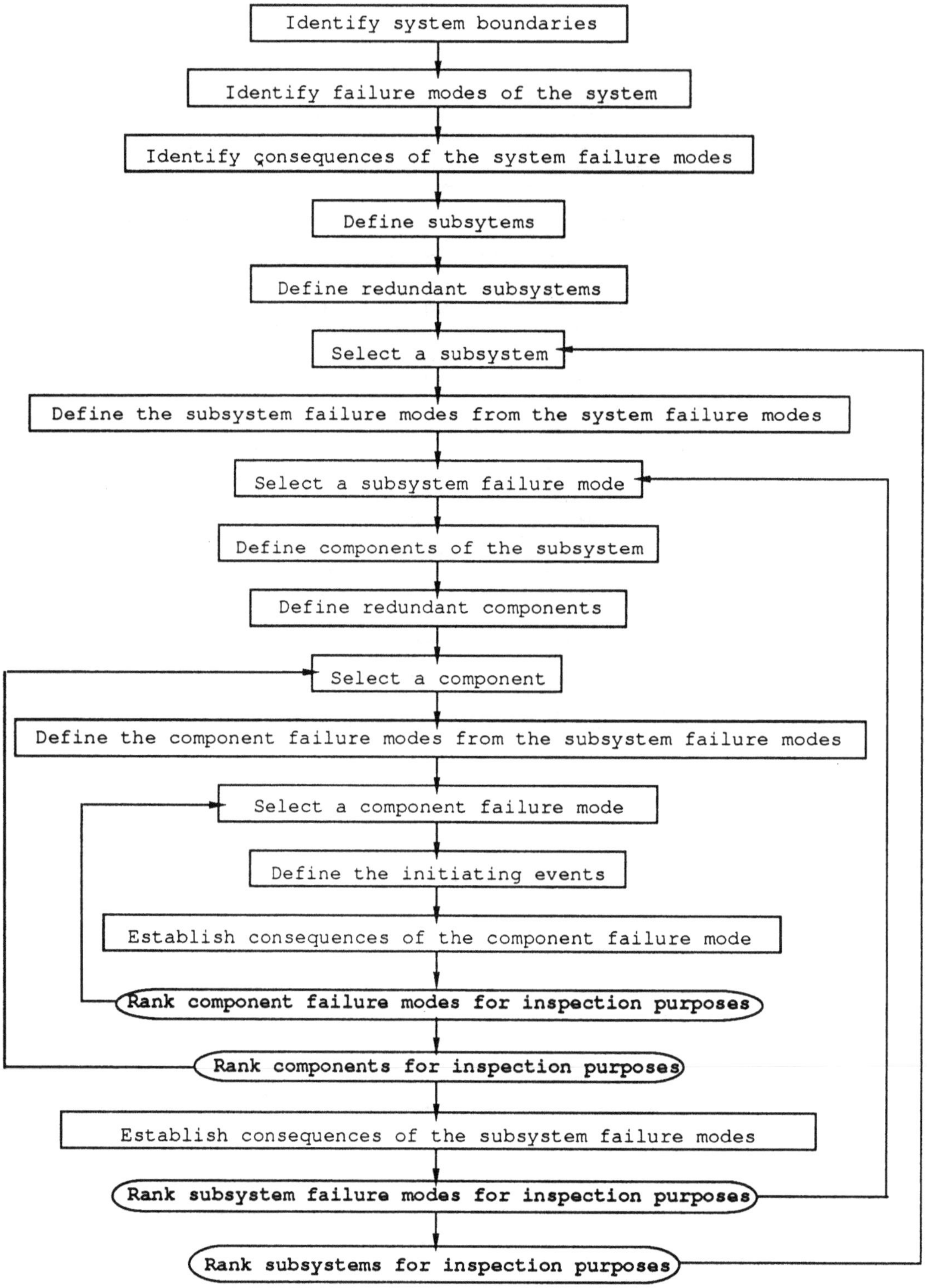

FIG. 2-3 TOP-DOWN QUALITATIVE ANALYSIS FOR INSPECTION RANKING AND CONSEQUENCE DEFINITION

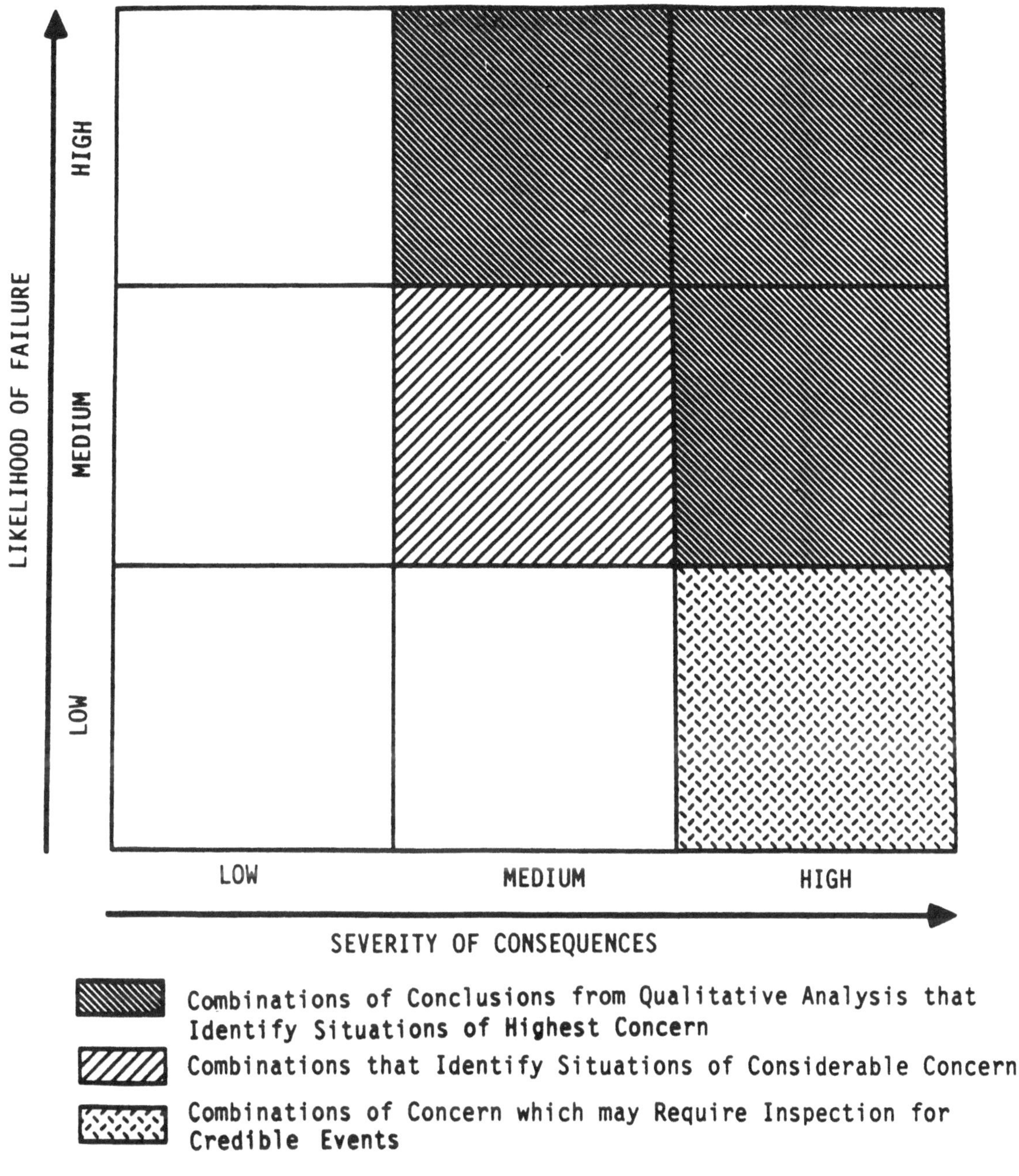

*Adapted from State of California's Guidance for the preparation of a Risk Management and Prevention Program by Lercari (1989), which was modified from EPA Technical Guidance for Hazards Analysis (1987).

FIG. 2-4 QUALITATIVE RISK-BASED RANKING MATRIX*

based process that will rank inspection priorities. On the other hand, judgment must be exercised to ensure that the evaluation is not unnecessarily diverted to consideration of large numbers of failure scenarios with very low likelihoods of failure.

Examples of simple structural failure modes that can lead to loss of structural integrity are:

- fracture
- excessive distortion
- cracking
- thinning

The consequences of such failure modes will be situation dependent. For example, a through-wall flaw that results in leakage of toxic or flammable fluids could have severe consequences. On the other hand, the consequences of this same structural integrity failure could be quite insignificant if a nonhazardous fluid is being contained.

Table 2-2 lists a number of important failure causes that should be considered in an assessment. Each of these can result in degradation of structural integrity. In the qualitative risk-assessment process, the most likely failure modes and causes must be identified based on experience and engineering judgment. The specific locations and components within the system or subsystem where the degradation mechanisms are expected must also be identified.

In many cases, this selection process is influenced by considerations of uncertainty. For example, when the details of the design and construction are either undocumented or suspect, or when there may be specific reasons to believe that components have been operated at temperatures, loads, or pressures beyond their original design limits, additional failure modes and causes may need to be considered. As a rule, the selection is based largely on general considerations of potential degradation from such causes as corrosion, cyclic stresses, wear, and high-temperature creep.

Since structural integrity failures are rare events, historical data on actual failures is likely to provide little guidance to identify the potential failures to be addressed by the risk-based inspection program. Therefore, engineering judgment is needed to first define the factors that can contribute to potential failures, and then to identify those factors that are of sufficient importance to warrant further consideration for the specific systems and components of concern.

2.3.4 □ Example

A generic example of how the qualitative risk-assessment process works, using the top-down approach, is illustrated by evaluating the failure of the hydraulic brake system in an automobile. The hydraulic brake system can be divided into two main subsystems, front and rear brake subsystems. The subsystems in this case are redundant such that in order for the automobile brake to fail, the two subsystems must fail simultaneously. Each subsystem can be divided into several components. The components can include the wheel cylinders, brake lining, master cylinder, connecting lines, and brake fluid. The first component (the wheel cylinder) can be divided into two subcomponents, i.e., left and right wheel cylinders. The failure of a subcomponent can cause the failure of cylinder function. Failure modes can include insufficient brake fluid, cut of the fluid lines, and seal failure in the master cylinder. Establishing consequences for each failure mode, component, and subsystem can be used as criteria for ranking them for inspection purposes.

This example could also be evaluated using the bottom-up quantitative analysis approach, but the process would begin with the definition of consequence of failure for the hydraulic brake system.

TABLE 2-2
EXAMPLE FAILURE CAUSES

- Operation at loads and/or pressures exceeding design limits
- Operation at temperatures over design limits
- Operation at temperatures below brittle fracture limits
- Improper design and fabrication
- Improper repairs and alterations
- Structural damage from maintenance
- Improper or degraded supports for components
- Structural damage from *external events* (impact, crushing, etc.)
- Excessive vibration
- Improper or degraded overpressure protection
- Material or weld defects
- General corrosion
- Flow-assisted corrosion (erosion/corrosion)
- Wear (excessive maintenance)
- Thermal fatigue cracking
- Vibrational fatigue cracking
- Stress corrosion cracking
- High-temperature creep
- Long-term embrittlement
- Loose or missing fasteners

2.4 □ QUANTITATIVE RISK ANALYSIS

Quantitative risk-analysis methods are used to integrate the numerous engineering disciplines to prioritize and develop programs for the inspection of components and structural elements. Some of the engineering disciplines include nondestructive examination, fracture mechanics analysis, probabilistic analysis, failure consequence analysis, system and component design, and operation of facilities.

Risk-based models are developed by expanding on the logic that is used in the qualitative risk-assessment process to quantify the direct and indirect consequences of failure. Probabilities of component failure that are based on component material, potential degradation mechanisms, and loading conditions are also factored into the model. Information from the design analysis and experience databases is also used in the risk-based prioritization model. The overall model is then used to identify the most important components or subcomponents for inspection using quantitative measures. A quantitative methodology, which can be based upon safety and/or economic risk, is then used to prioritize all the components or elements of interest. When the components are prioritized, inspection models are applied to evaluate appropriate inspection strategies for the components or structural elements of interest. These models are used to evaluate the reliability of inspection techniques and inspectors relative to potential failure criteria, e.g., leaks or catastrophic rupture.

The overall approach is essentially comprised of two basic steps:

(1) the application of risk-analysis methods to prioritize components for inspection; and

(2) the application of decision risk-analysis models to evaluate the appropriate frequency, methods, and acceptance criteria for developing inspection programs.

Section 2.5 discusses enhanced failure modes, effects, and criticality analysis (FMECA) model that has been altered to develop component inspection priorities. Decision analysis models for developing inspection programs are discussed in Section 2.6.

The FMECA model has been expanded beyond traditional FMECA models used for evaluating system effects from operational component failures. This enhanced FMECA methodology illustrates in a straightforward way how risk is used to prioritize components for structural inspection. Rather than using qualitatitve values or weighting factors that are applied in traditional FMECA models, numerical failure probabilities and consequences are assigned in this enhanced method. Approaches for treating uncertainty in the assigned values are also offered.

Strictly speaking, the FMECA approach is only appropriate where single failures are linked directly to specific consequences. Where combinations of independent failures are required to cause the undesirable consequences, this method must be enhanced or replaced by more sophisticated analysis methods. In that case, the probability of component failure must be combined with the probabilities of other failures to determine the probability of achieving the consequences.

Thus, where safety systems exist to respond to component failures, and in particular where independent, redundant systems exist, fault trees may be used to quantify probabilities. Where time-sequencing of events must be considered, i.e., *accident sequence,* event trees may be used (often in combination with fault trees) in probability determinations. Both of these techniques are used in probabilistic risk assessments (PRA), which are extensively used in the nuclear industry for making probability determinations where these complications exist.

Where PRA information exists, more sophisticated measures of *risk importance* may be used to prioritize component inspections, as is discussed in Section 2.6.2. Nevertheless, the FMECA model is appropriate for use in less complicated situations, and it is particularly appropriate for the purposes of this document – to provide a straightforward illustration of how risk can be used to prioritize components for inspection.

2.5 □ FAILURE MODES, EFFECTS, AND CRITICALITY ANALYSIS METHODOLOGY

2.5.1 □ FMECA Information Sources

The FMECA methodology provides an efficient way to integrate the information that is required to perform a risk-based prioritization. Figure 2-5 depicts how several sources of information are used to construct the FMECA model. The following paragraphs summarize each of these inputs.

Design Information. Design information is used to identify the subsystems and components within a given structural system. This information is pertinent to defining functional requirements and expected loadings for components and structures. Insights can be obtained as to definition of the potential failure modes and causes from this data. For instance, locations in piping systems that are susceptible to fatigue or erosion can be determined by noting those locations where frequent thermal transients or high flow rates occur, respectively.

Operating Experience Databases. An excellent source for determining single and multiple failure modes and failure causes for some components and structures is operating experience data. [For example, Bush (1988) provides useful data on pressure vessel and piping failures.] For many components and structures, however, a dearth of data exists on actual service-induced failures. Engineering analysis should be used to extrapolate data so failure probabilities for these components can be assigned, particularly when estimating long service times, as needed in life-extension evaluations. Inferences can be drawn from these sources to define general trends.

Expert Opinion. Elicitation of expert opinion of the personnel who have first-hand field experience from operating and maintaining the facility of interest is another valuable source. These persons usually know of failures, malfunctions, degradation, and *human errors* that are not documented, but are important to the failure probability estimation. Winterfeldt and Edwards (1986) provide general methods for eliciting expert opinion or judgment. If expert opinion is being sought in order to encode probabilities for analytical purposes, methods such as those presented by Spetzler and Von Holstein (1975) must be applied. Many pitfalls exist when eliciting expert opinion for analytical purposes. Mosleh, Bier, and Apostolakis (1988) highlight some of those concerns in their review of current practice for use of expert opinion in probabilistic risk assessment. Use of a well-documented systematic process that is facilitated by experienced practitioners to obtain this valuable source of information avoids wasting time and resources.

Potential Degradation Mechanisms. Studies of potential degradation mechanisms can help to define potential failure modes and causes, which were previously discussed in Section 2.3. Such studies have been performed in support of life-extension efforts and can provide useful insights into predicting failures that have not yet occurred in operation. Once again, the knowledge of personnel with field experience can often be used to identify undocumented degradation mechanisms.

Prior Inspection Results. Review of prior inspection results further supplements the above information when attempting to define the potential for component failure. The experience range, from finding nothing to having to repair or replace the component of interest because of significant flaw indications, should be taken into account. Flaw indications exceeding acceptance standards that have been found in comparable components may usefully supplement these results. Given the void in databases for most components and structures, this source of data can have a significant influence in assigning probabilities of failure for the components or elements of interest.

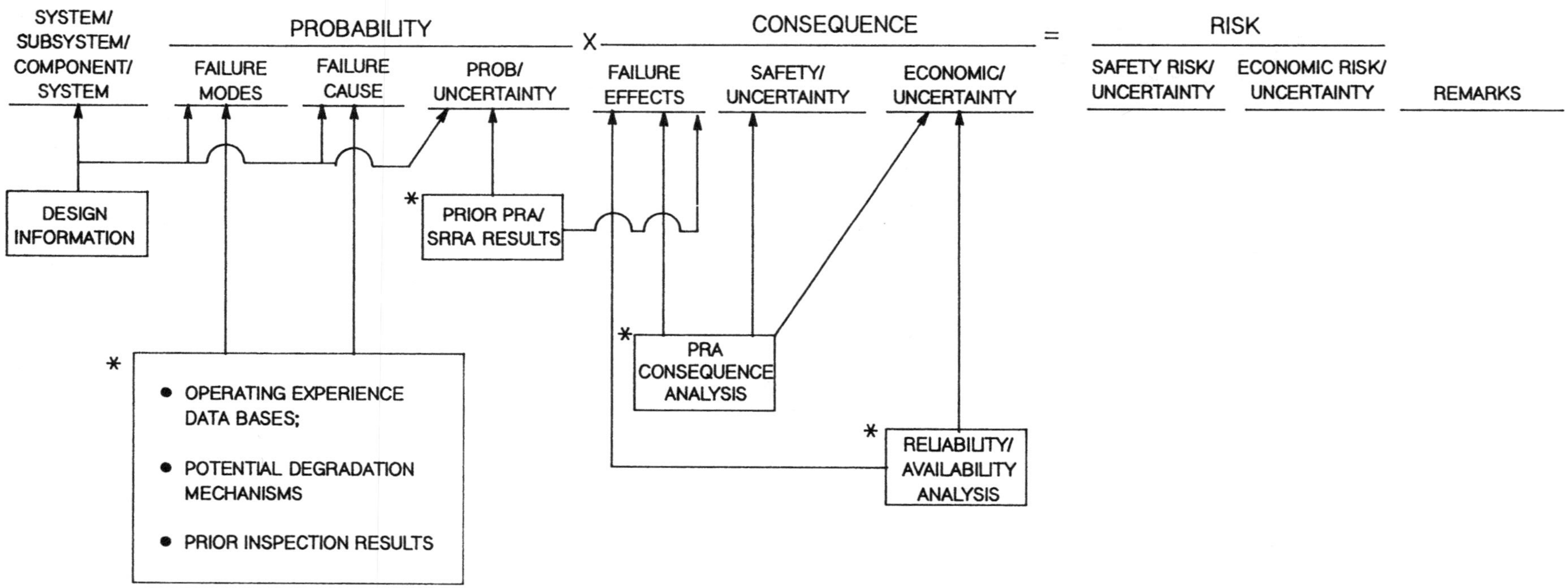

* EXPERT OPINION CAN ALSO BE USED HERE

FIG. 2-5 INTEGRATION OF TECHNICAL INFORMATION INTO FMECA FOR RISK-BASED INSPECTION

SRRA Results. Prior *structural reliability and risk assessment (SRRA)* methods can be used to integrate the above information when predicting the probability of failure for a given component, particularly where a dearth of failure data exists. These methods address the uncertainties associated with flaws, material properties, and loadings. They can also be combined with traditional probabilistic risk assessment (PRA) results that define the frequency of loading events on components and structures. SRRA results have been generated for a number of components over the past decade and can be of use in assigning failure probabilities. (In Section 2.6 this technology is further discussed in detail, relative to the development of inspection strategies for components.)

PRA and Reliability Analysis Results. Available PRA consequence and reliability/availability systems analysis can assist in determining effects of component failures for the consequence portion of the FMECA. PRA evaluations provide useful data relative to safety consequences, such as the potential for damage and undesirable releases. Systems reliability analyses provide useful information regarding unplanned shutdowns as a result of failures, particularly those that may result in extended facility shutdown with minimal safety consequences. These tools provide a good source for the FMECA; however, one should include uncertainty in the results since these analysis methods do not currently include time-related degradation.

2.5.2 □ Estimation of Failure Probabilities (Prob)[1]

To estimate failure probabilities for the components or elements of interest for the FMECA, appropriate failure data must be used in combination with available SRRA and/or PRA results. For most applications, however, this information does not exist and the estimation has to be made by engineering judgment using the other sources discussed above.

One of the challenges with estimating failure probabilities is converting statements in the spoken word to actual probabilities. For example, suppose that a facility, which has a lifetime measured in tens of years and where a particular event may be expected to occur during the lifetime of the facility, then what probability rate (i.e., frequency) would be applied to that event? Or suppose that another event is rarely expected, but could not be ruled out completely, then what probability rate would be applied to this event? Table 2-3 gives some definitions that have been generally agreed to be equivalent to these types of statements. This table can help in those situations where the data is scarce or nonexistent and is relied upon to make judgments using these types of statements. However, undertaking these translations from the spoken word of an expert to a probability value is a process with pitfalls and should be approached most carefully, as previously discussed in the elicitation of expert opinion in Section 2.5.1.

The probabilities provided in Table 2-3 are presented on a per-year basis and are applicable to components whose lifetimes are measured in tens of years. These numbers must be adjusted for short-lived components, such as rocket engines whose lifetimes are measured in minutes. Another consideration when quoting probabilities is that some situations call for a per-demand basis. This would apply to a component, such as a fire pump, that is seldom in use, but which becomes very important when the situation calling for its use arises.

---

[1] Abbreviation used in FMECA output.

TABLE 2-3
DEFINITION OF FAILURE PROBABILITY ESTIMATES (PROB) FOR COMPONENTS WITH LIFETIMES IN TENS OF YEARS

Definition	Failure Probability (per year)
An off-normal initiating event which individually may be expected to occur more than once during the lifetime of the component.	10^{-1}
An off-normal initiating event which individually may be expected to occur once during the lifetime of the component.	10^{-2}
An off-normal initiating event which individually is not expected to occur during the component lifetime; (however, when integrated over all system components, an event in this category has the credibility of happening once).	10^{-4}
An off-normal initiating event of such low probability that an event in this category is rarely expected to occur.	10^{-6}
An off-normal initiating event of such extremely low probability that an event in this category is considered to be incredible.	10^{-8}

2.5.3 □ Consequence Estimation

The consequence portion of the FMECA should be represented by both safety and economic consequences given component failure. Environmental consequences can be included in either of these categories, or perhaps addressed separately. Safety effects should consider direct as well as indirect consequences since these effects are generally not considered in PRA evaluations from a structural integrity point of view. Otherwise, available PRA and systems availability analysis results are the best source for the assignment of the consequence estimate. No generally accepted method exists for combining safety and economic consequences to perform a risk-based ranking or prioritization for inspection. Therefore, it is recommended that they be considered separately as appropriate to the application of interest.

Safety. For the general methodology, the safety consequence should be defined in terms of the potential for adverse effects (e.g., the number of casualties, both on-site and off-site, that would be expected to occur given the failure of the component or element of interest, particularly for a *significant release*). Safety consequences can be defined for specific applications, e.g., core damage for nuclear power plants or size of toxic release clouds for process facilities, but these will be specified in later volumes for the application of interest. The safety consequences can sometimes be obtained from PRA consequence evaluations, such as *dispersion analysis* and *release consequence analysis.* Once again, however, this data is generally not readily available. Therefore, the safety consequence estimate is made based on engineering judgment for most cases.

Economic Loss. The consequence of economic loss will have to be estimated, but experience usually exists from other related failures that can be drawn upon for the estimation, even for very serious accidents. Economic loss entails the repair or replacement of the component or element that has failed, the repair or replacement of other equipment that was affected by the failure, the loss of availability of the overall

TABLE 2-4
EXAMPLE VALUES FOR ECONOMIC LOSS CONSEQUENCES

Definition	Estimated Economic Loss, $
Failure causes significant potential off-site and facility or system failure costs and potential for significant litigation	10^9
Failure causes indefinite shutdown, significant facility or system failure costs, and potential for litigation	10^8
Failure causes extended unscheduled loss of facility or system and significant component failure costs	10^7
Repair can be deferred until scheduled shutdown, some component failure costs will occur	10^5
Insignificant effect on operation	10^4

system or facility, business interruption, and in some cases, damage to the public or environment. Some suggested economic loss values, which can be refined using expert opinion for the application of interest, are presented in Table 2-4.

A more universal way can be established to quantify the consequences of property damage using a units-of-damage measure reflecting well-defined components of a system and reflecting parameters that modify the cost of such consequences. Among these parameters one can include the following:

(1) transportation;

(2) installation, including special technology methods and resources needed;

(3) modification of linked components of the system because of new technology advancement of the object being replaced;

(4) cost of expediting the repair or replacement;

(5) business interruption; and

(6) legal and jurisdictional expenses.

The basis of the evaluation of the total cost could be the direct replacement cost of the system (vendors price list) equated to 100 units. Any partial damage of the system would be linked to a cost of a number of units (less than 100) modified with the number of units reflecting the above parameters applicable in the site-specific case. Total damage would reflect a cost of at least 100 units.

The advantage of this method is the independence of damage cost prediction from currency changes (inflation, exchange rates for foreign-made equipment, etc.) and technological advancements such as new processes or new equipment. The disadvantage is the need for normalization of the damage units if the cost of the damage is the basis for comparison and ranking of systems.

2.5.4 □ Risk Evaluation

As previously shown in Fig. 2-5, the risk of failure is obtained by multiplying the probability of failure by the consequence of that failure for each component or structural element of interest. To perform the prioritization for inspection, each component or element can be ranked from highest to lowest according to its safety or economic risk. A useful way to evaluate the risk of failure, however, is to develop risk plots in addition to just preparing a tabulated ranking of the risk.

The probability and consequence of failure for each component or element in a given system are plotted on a log–log risk plot using single point estimates. One plot is made for safety risk and another for economic risk, depending on the interest of the user. Ranking can be done via lines of constant probability of failure (Fig. 2-6a), lines of constant consequence (Fig. 2-6b), or lines of constant risk (Fig. 2-6c). Generalized contours can also be drawn to place emphasis on failure probability or consequence when ranking or grouping components (Fig. 2-6d). Special cases of the generalized plots can be generated such as for insurance evaluations (Fig. 2-6e). Typically, inspection of components or elements associated with large consequence are performed to verify that some unknown degradation mechanism is not taking place.

Groupings are used to classify or rank components. For example, all the components or elements in Category "A" may receive full examination at a frequent inspection interval using the best inspection methods available. SRRA models may also be developed for these elements in order to define the optimum inspection strategy and to assure correct categorization. (This is discussed in more detail in the next section.) In Category "B," samples may be used for the inspection, but the frequency and methods of inspection may be similar to those in Category "A." Finally, in Category "C," only visual inspections may be performed on a routine basis to ensure that major degradation or minor failure (e.g., small leaks or bulging) is not taking place.

This document does not define "acceptable risk" levels since it is up to the user of this technology (i.e., the owner, vendor, regulatory body, insurer, etc.) to define those risk levels for the application of interest.

Figures 2-6a through 2-6e use *single point estimates* to evaluate risk. In discussing the input to the FMECA, significant uncertainty can exist, however, within the probabilities and consequences of failure as previously mentioned. The quantitative risk analysis may be carried to a higher level by treating uncertainties, as is discussed and exemplified next.

2.5.5 □ Treatment of Uncertainties

The main motivation behind the evaluation of safety and risk in engineering is to deal with uncertainties for the purpose of risk assessment and risk control. Therefore, a comprehensive and complete treatment of uncertainties in engineering systems is needed in order to obtain realistic estimates of risk. Generally, uncertainties concerning engineering systems are commonly attributed to:

(1) physical randomness in system parameters;

(2) statistical uncertainty due to the use of limited information to estimate the characteristics of these parameters;

(3) model uncertainties that are due to simplifying assumptions in our analytical and prediction models, simplified methods, and idealized representations of reality;

(4) vagueness in the definition of certain parameters, e.g., system performance (failure or survival), quality, deterioration, skill and experience of operators and engineers; and

(5) ambiguity and vagueness in defining the relationships between the parameters of the problems, especially for complex systems.

Figures 2-7a through 2-7e are replots of the data in Figs. 2-6a through 2-6e, taking uncertainty into account using *interval estimates* on the probability of failure and the consequence. That is, the probability and consequence data is represented by rectangles rather than single points to display the risk and inherent uncertainty of each element. Risk rankings can be established to prioritize components for inspection; however, elements with large uncertainty (i.e., large or stretched rectangles) will be

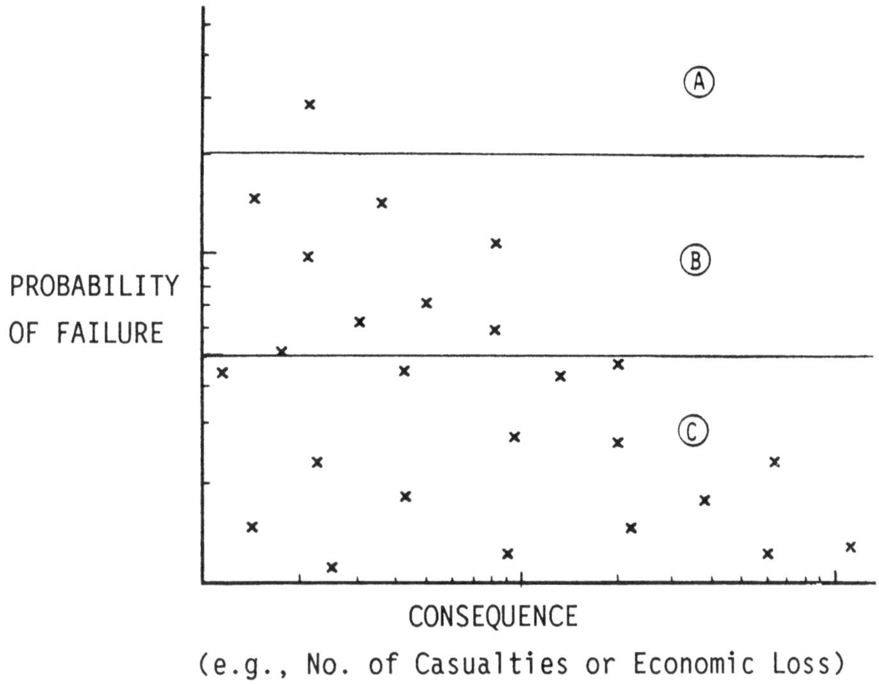

a. Ranking Based Only on Probability of Failure

b. Ranking Based Only on Consequence

FIG. 2-6 RISK RANKING USING POINT ESTIMATES

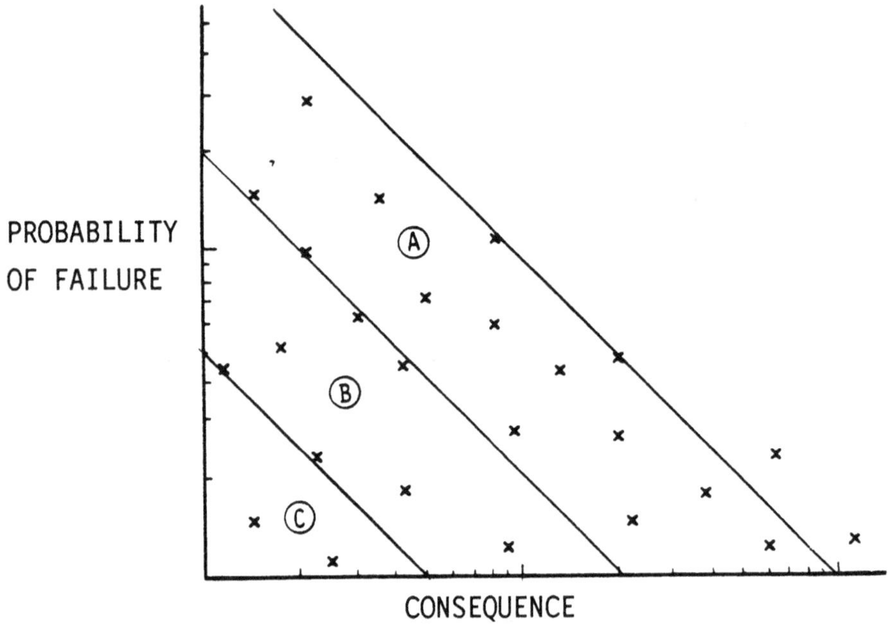

c. Ranking Based on Lines of Constant Risk

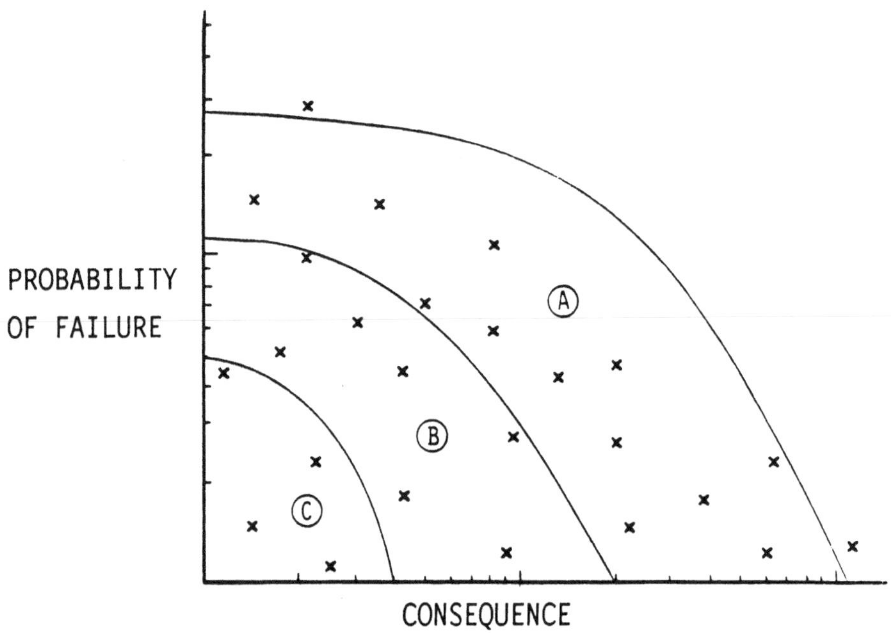

d. Ranking Based on Generalized Risk Contour Lines

FIG. 2-6 RISK RANKING USING POINT ESTIMATES (CONT'D)

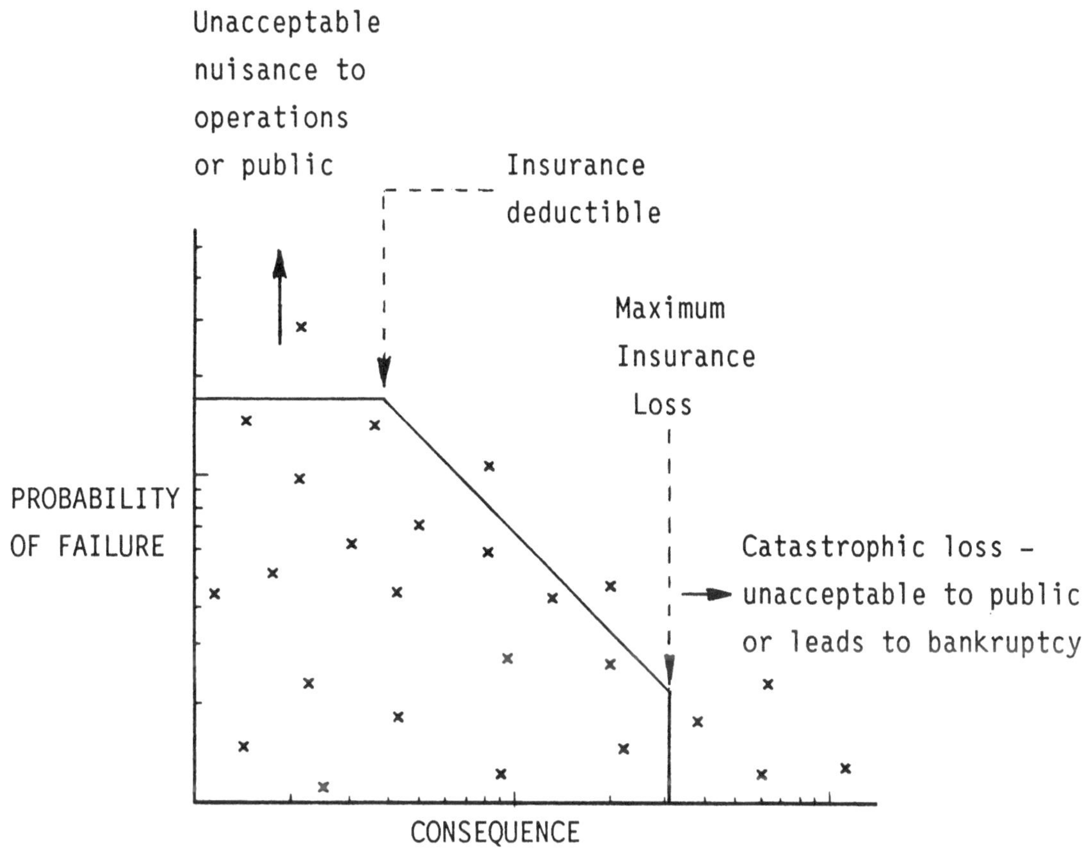

e. Ranking Based on Special Case of the Generalized Methodology for Insurance Evaluations

FIG. 2-6 RISK RANKING USING POINT ESTIMATES (CONT'D)

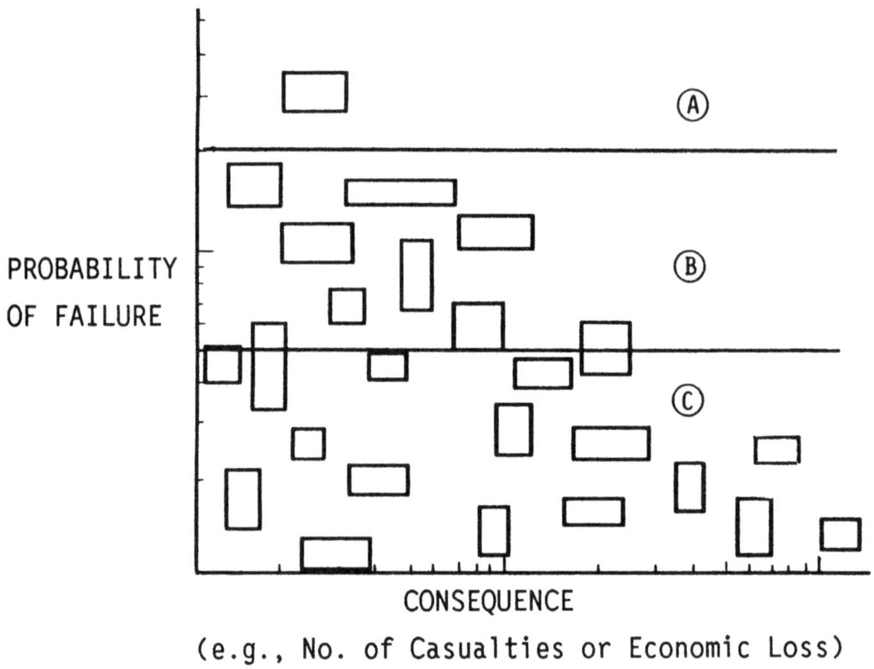

a. Ranking Based Only on Probability of Failure

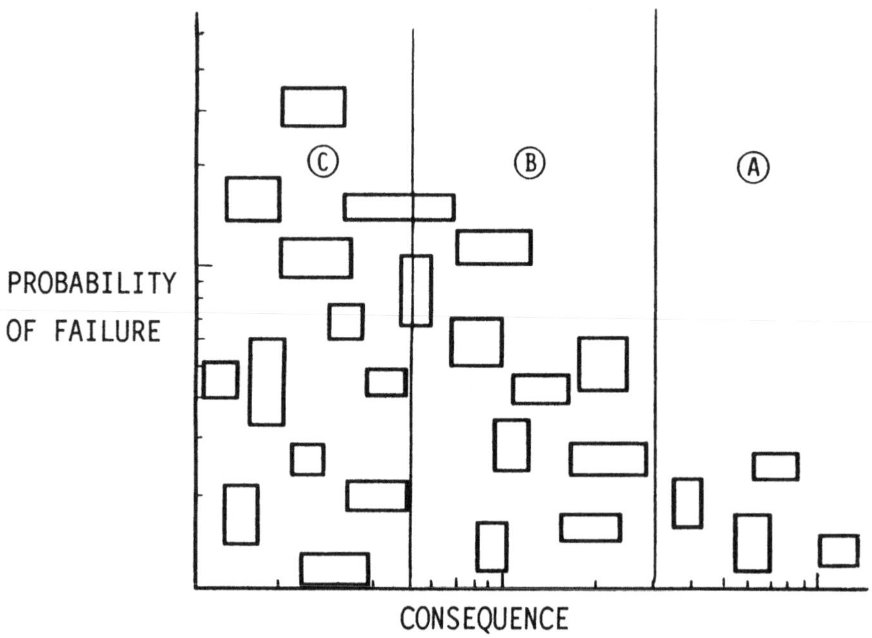

b. Ranking Based Only on Consequence

FIG. 2-7 RISK RANKING INCLUDING UNCERTAINTY

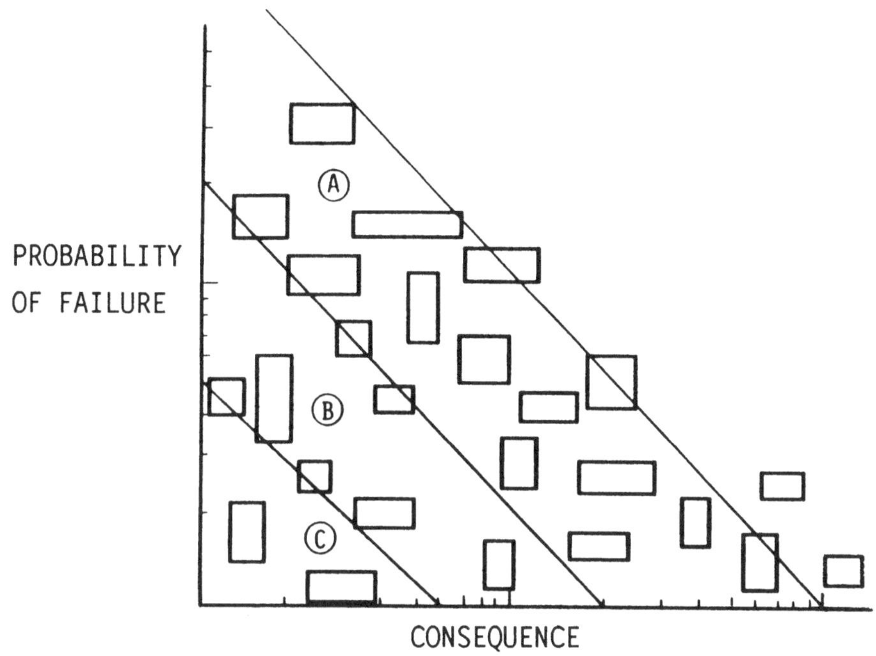

c. Ranking Based on Lines of Constant Risk

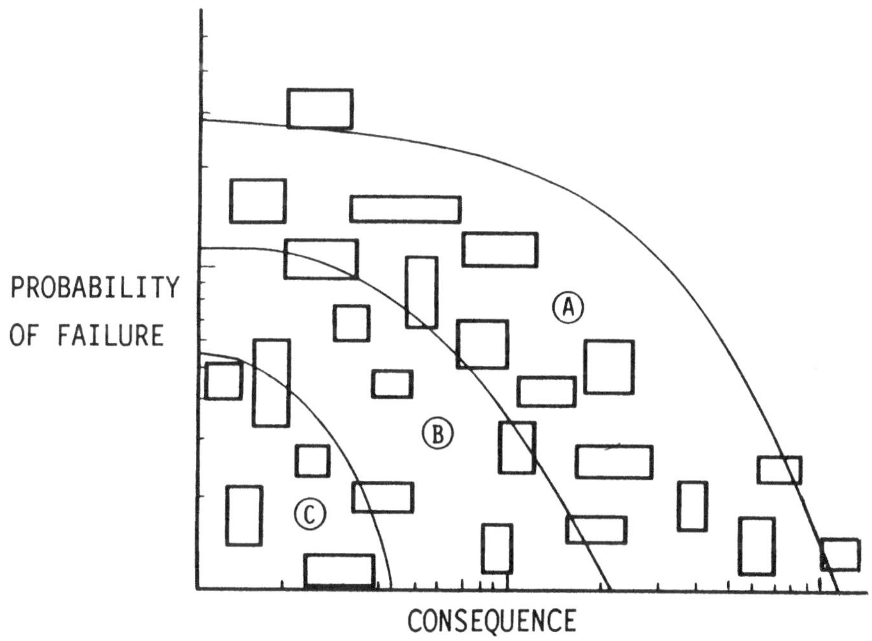

d. Ranking Based on Generalized Risk Contour Lines

FIG. 2-7 RISK RANKING INCLUDING UNCERTAINTY (CONT'D)

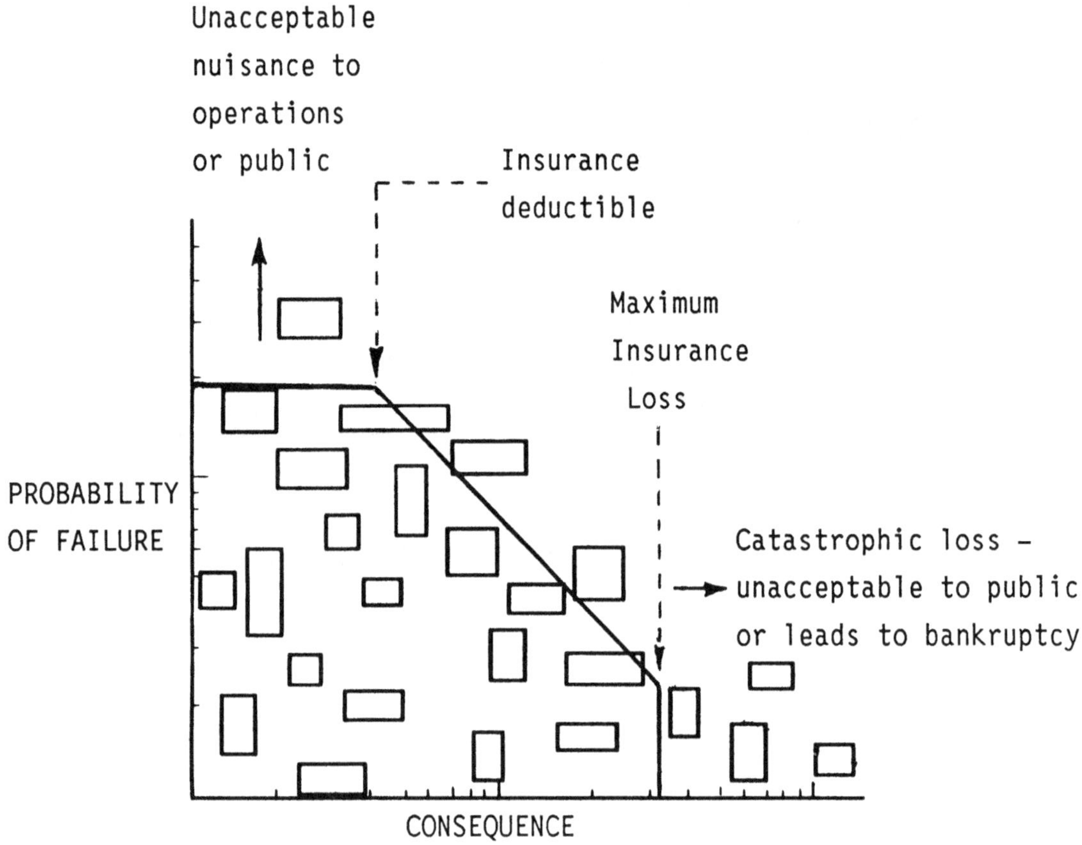

e. Ranking Based on Special Case of the Generalized Methodology for Insurance Evaluations

FIG. 2-7 RISK RANKING INCLUDING UNCERTAINTY (CONT'D)

ranked higher than in the grouping based on single point estimates. Elements, represented with rectangles, with risk estimates that fall at borderlines between groups can be ranked in a group of higher inspection priority.

The next section discusses the methodology and provides examples using interval estimates as a recommended way to quantify uncertainty if the user prefers to go beyond the use of single point estimates for the risk-based inspection process. More complex uncertainty measures are available in the literature.

Uncertainty measures discussed here are classified into the following three major caterogies:

(1) information measures based on crisp sets and probability theory;

(2) uncertainty and information measures based on fuzzy sets and possibility theory; and

(3) uncertainty measures within the framework of belief and plausibility measures, i.e., the theory of evidence.

In the first category, there are two principal measures of uncertainty: the Hartley measure, which is based solely on the classical set theory, and the Shannon entropy, which is formulated in terms of probability theory. In the second category, there are two principal measures of uncertainty: the U-uncertainty, which is a possibilistic counterpart of Shannon entropy and, at the same time, a generalization of the Hartley measure, and the measure of fuzziness, which is based on fuzzy set theory. The third category consists of dissonance and confusion measures that are based on the theory of evidence. These uncertainty measures are summarized in Table 2-5 for information.

2.5.6 □ Risk-Based Ranking Methodology

Based on the criteria set by the task force, a methodology for risk-based ranking of components or elements for inspection is developed. The methodology is based on the assessment of failure modes, element failure probabilities, consequences, and the uncertainties associated with them. Single and multiple element failure modes are considered since, for some elements, several failure modes may exist. Safety and economic consequences are included. As part of the uncertainty analysis, interval risk estimates are considered for failure modes, elements, and systems.

Several additional examples are used to illustrate the suggested methodology for ranking failure modes, elements, and systems for inspection purposes. Two approaches are considered in this analysis. They include: single point estimate of risk and interval estimate of risk. The selection of the method depends on the application and the availability of information.

The objective of the analysis is to construct an inspection priority list. The priority list is determined using safety and economic risks for all the failure modes, elements, and subsystems of the system under investigation. Generally, for both single and multiple failure modes, the ranking is based on the severity potential of safety and economic risks. In the case of tied rankings, the statistically appropriate approach for determining the ranking of the tied elements is to average the ranks of these elements. In this section, however, the ranking of tied elements is taken as the least rank among them. The next available rank is then taken as the least rank plus the number of tied elements. This practice is commonly used in the development of ranks.

Single Point Estimate. In the single point estimate, values are assigned to the probabilities of failure of the modes and elements, and the magnitudes of safety and economic consequences. The resulting estimates of risk are point values. The safety and economic risks are the products of the probabilities of failure and the corresponding magnitudes of safety and economic consequence, respectively. Ranking of the safety and economic risk estimates for the elements and systems are performed from highest to lowest.

TABLE 2-5
UNCERTAINTY MEASURES

	A	B	C	D	E	F	G
1	Uncertainty measure	Type of uncertainty	Type of sets or events	Theory type	Comments	Uncertainty range	Reference
2					A basic discrete measure.		
3	Hartley	ambiguity	crisp	set	A larger number of outcomes	$[0,\infty)$	Hartley [1928]
4					means larger uncertainty		
5				set	The closer the outcomes		
6	Shannon Entropy	ambiguity	crisp	and	to an equal liklihood, the	$[0,\infty)$	Shannon [1948]
7				probability	larger the uncertainty		
8				set	Possibilistic counterpart		
9	U-uncertainty	ambiguity	crisp	and	to Shannon entropy and	$[0,\infty)$	Higashi and Klir [1983]
10				possibility	generalization to Hartley		
11					measure		
12		vagueness		set	Measures the lack of		Deluca and Termini
13	Fuzziness measure	and	fuzzy	and	distinction between a set	$[0,\infty)$	[1972,1974,1977]
14		ambiguity		fuzziness	and its complement		
15		conflict		set	Measures conflict of		
16	Dissonance measure	and	crisp	and	evidence using theory of	$[0,\infty)$	Yager [1983]
17		ambiguity		evidence	evidenc		
18		confusion		set	Measures confusion of		
19	Confusion measure	and	crisp	and	evidence using theory of	$[0,\infty)$	Hohle [1981]
20		ambiguity		evidence	evidenc		

Interval Estimate. An interval value is a range of numbers with lower and upper limits that is assigned to some or all the parameters considered above to account for the uncertainty in these parameters. For example,

$$I = [a,b] \tag{2-1}$$

is an interval value ranging from a lower limit *(a)* to an upper limit *(b)*. The numbers in the interval, i.e., from *a* to *b*, can have either a uniform or non-uniform likelihood distribution as shown in Fig. 2-8. In this study, it is assumed that the numbers of an interval have uniform likelihood distributions.

The operations of interval analysis can be used to determine uncertainty propagation in the analytical process. The safety and economic risks are the products of the interval probability of failure and the intervals of the magnitudes of safety and economic consequence, respectively.

The algebraic operations on interval values are extensions of operations on real numbers as discussed by Moore (1966). For example, if

$$I_1 = [a,b] \tag{2-2}$$

and

$$I_2 = [c,d] \tag{2-3}$$

are two interval values where $d > c$ and $b > a$, then the following operations are defined:

$$[a,b] \times 0 = 0 \tag{2-4}$$

$$[a,b] + [c,d] = [a + c, b + d] \tag{2-5}$$

$$[a,b] - [c,d] = [a - d, b - c] \tag{2-6}$$

$$[a,b] \times [c,d] = [ac,bd] \tag{2-7}$$

$$[a,b]/[c,d] = [a,b] \times [1/d,1/c] \text{ if 0 does not belong to } [c,d] \tag{2-8}$$

The distribution shape of the resulting interval from any operation on two intervals can vary from the original distribution shapes. The resulting shape depends on the shapes and the magnitudes of the original intervals. For example, the addition of two uniformly distributed intervals results in a non-uniform interval. The distribution types are not considered as part of the analysis in this study.

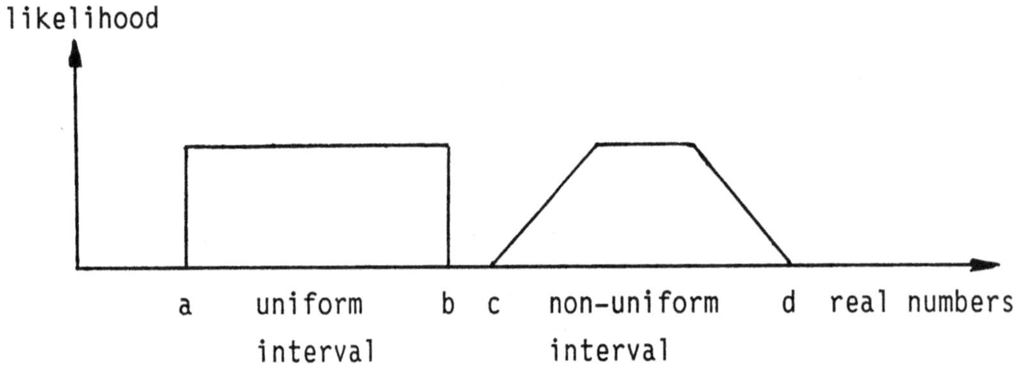

FIG. 2-8 LIKELIHOOD DISTRIBUTIONS

In order to rank the safety or economic risk estimates where their magnitudes are given by interval values, for example

$$I_1 = [a,b] \tag{2-9}$$

and

$$I_2 = [c,d] \tag{2-10}$$

the following logic operations should be conducted:

$$\text{IF } d > b \text{ THEN } I_2 > I_1 \tag{2-11}$$

$$\text{IF } d = b \text{ AND IF } c > a \text{ THEN } I_2 > I_1 \tag{2-12}$$

$$\text{IF } d = b \text{ AND } c = a \text{ THEN } I_2 = I_1 \tag{2-13}$$

$$\text{Otherwise } I_2 < I_1 \tag{2-14}$$

2.5.7 □ Risk-Based Ranking Examples

In this section, two simplified examples are discussed to illustrate the process of risk-based ranking. The first example is based on elements with single failure modes. Both single point and interval estimation methods are used to illustrate the methodology. In the second example, multiple failure modes for elements are considered using both estimation methods. The elements are assumed to be in series for both examples, and the probability of more than one element failure is assumed to be negligible.

Prior to showing these numerical calculations, Table 2-6 illustrates the basis for the failure probability and consequence estimates that are used in the FMECA of the general structural system being evaluated in the two examples. The consequence estimates for the examples are based on the potential number of casualties and economic

TABLE 2-6
FMECA RISK-BASED INSPECTION OF GENERAL STRUCTURAL SYSTEM

SYSTEM	ELEMENT	FAILURE MODES	FAILURE CAUSE	PROBABILITY OF FAILURE	FAILURE EFFECTS	CASUALTIES	DAMAGE	REMARKS
1	1	Distortion that leads to weakening of redundant member	Excessive temperatures - significant dynamic event, e.g., large earthquake	1.00E-14	Localized damage, no effect on public or operation	0.00E+00	0.00E+00	None
	2	Fracture that results in leakage of hazardous material	Corrosion, mechanical wear, poor welding process	1.00E-14	Potential threat to drinking supply in local area, minor effect on operations	1.00E+03	1.00E+03	None
	3	Buckling that leads to total collapse of system	Erosion - significant dynamic event	1.00E-14	Major release of toxic material to population within radius of 20 miles, loss of facility and operations	1.00E+06	1.00E+06	Redundant member could be easily installed to further prevent the failure from ever occurring
	4							
	.							
	.							
	.							
	7	Fracture that results in minor leakage	Corrosion, mechanical wear, vibrational fatigue-cracking from nearby pump	1.00E-02	No effect on public, negligible effect on operation	0.00E+00	0.00E+00	Significant cracking has been observed on previous inspections of similar systems
	.							
	.							
	.							

loss or damage from the potential failures that are considered to obtain human and economic risk, respectively. These consequence definitions are used only for the purposes of illustration.

Example 1: Single Failure Modes

Single Point Estimate. Single point estimates for the probabilities of element failure, casualties, and damage are assumed as shown in Table 2-7, columns C, D, and E, respectively. Table 2-7 shows two systems with nine elements each. The probabilities of element failure are selected to range from 1.00E-14 to 1.00E-02. The magnitudes of casualty are selected to range from 0 to 1 million in some units. The magnitudes of damage are selected to range from 0 to 1 million in some units. The human risk (column F), due to an element failure, is the product of the probability of element failure (column C) and the magnitude of casualty (column D). The economic risk (column G), due to an element failure, is the product of the probability of element failure (column C) and the magnitude of damage (column E).

The probability of failure of the system is bounded by lower and upper limits. The lower and upper limits correspond to statistically correlated (positive) and statistically independent failure events of the element, respectively. The limits are based on the assumption that the elements are in series. The limits for the probability of system failure can be determined as follows:

$$\operatorname*{Max}_{i=1}^{i=9} P_{fci} < P_{fs} < 1 - \pi_{i=1}^{i=9}(1 - P_{fci}) \tag{2-15}$$

where

P_{fci} = the probability of failure of element i
P_{fs} = the probability of system failure

The estimated probability of system failure can be assumed to be the geometric average of the two system failure limits, where the geometric average of Y_1, Y_2,....,Y_n is defined as the nth root of the product of all Y's, i.e.,

$$\text{geometric average} = (Y_1 \, Y_2 ... Y_n)^{1/n} \tag{2-16}$$

The risk estimates for a system are determined as the summation of all the risks of its elements as shown in Table 2-7. The human risk values are shown in cells 15-F and 28-F for systems 1 and 2, respectively. The economic risk values are shown in cells 15-G and 28-G for systems 1 and 2, respectively. The elements and systems can be ranked according to human and economic risks as shown in columns H and I, respectively. For system 1, the magnitude of casualties for each element is set equal to the damage magnitude in order to illustrate the effect of interval analysis as shown in the next section.

Interval Analysis. In order to perform an interval analysis for Example 1, the probabilities of element failure, and the magnitudes of casualty and damage are expressed in the form of intervals as shown in Table 2-8. Table 2-8 is similiarly constructed to Table 2-7 in order to facilitate a comparison between the two solutions as discussed in the next section. The resulting estimates of human and economic risk are expressed in the form of intervals. The mathematical operations that are used for this purpose are given by Eqs. 2-4 through 2-8.

TABLE 2-7
POINT ESTIMATE METHOD FOR SINGLE FAILURE MODES

	A	B	C	D	E	F	G	H	I
1	System	Element	Probability	Magnitude of				Ranking according to	
2			of Faliure	Casualties	Damage	Human Risk	Economic Risk	Human Risk	Economic Risk
3	1								
4		1	1.00E-14	0.00E+00	0.00E+00	0.00E+00	0.00E+00	8	8
5		2	1.00E-14	1.00E+03	1.00E+03	1.00E-11	1.00E-11	7	7
6		3	1.00E-14	1.00E+06	1.00E+06	1.00E-08	1.00E-08	6	6
7		4	1.00E-05	1.00E+00	1.00E+00	1.00E-05	1.00E-05	5	5
8		5	1.00E-05	1.00E+03	1.00E+03	1.00E-02	1.00E-02	4	4
9		6	1.00E-05	1.00E+06	1.00E+06	1.00E+01	1.00E+01	2	2
10		7	1.00E-02	0.00E+00	0.00E+00	0.00E+00	0.00E+00	8	8
11		8	1.00E-02	1.00E+03	1.00E+03	1.00E+01	1.00E+01	2	2
12		9	1.00E-02	1.00E+06	1.00E+06	1.00E+04	1.00E+04	1	1
13	Upper Limit		2.97E-02						
14	Lower Limit		1.00E-02						
15	Average		1.72E-02			1.00E+04	1.00E+04	1	1
16	2								
17		1	1.00E-10	0.00E+00	1.00E+03	0.00E+00	1.00E-07	8	7
18		2	1.00E-07	1.00E+03	1.00E+02	1.00E-04	1.00E-05	4	4
19		3	1.00E-08	1.00E+06	5.00E+03	1.00E-02	5.00E-05	2	3
20		4	1.00E-08	1.00E+00	1.00E+03	1.00E-08	1.00E-05	7	4
21		5	1.00E-07	1.00E+03	1.00E+00	1.00E-04	1.00E-07	4	7
22		6	1.00E-03	1.00E+06	0.00E+00	1.00E+03	0.00E+00	1	9
23		7	1.00E-06	0.00E+00	1.00E+06	0.00E+00	1.00E+00	8	1
24		8	1.00E-08	1.00E+03	1.00E+06	1.00E-05	1.00E-02	6	2
25		9	1.00E-09	1.00E+06	1.00E+03	1.00E-03	1.00E-06	3	6
26	Upper Limit		1.00E-03						
27	Lower Limit		1.00E-03						
28	Average		1.00E-03			1.00E+03	1.01E+00	2	2

TABLE 2-8
INTERVAL ANALYSIS FOR SINGLE FAILURE MODES

	A	B	C	D	E	F	G	H	I
1	System	Element	Probability	of	Failure	Magnitude	of	Casualties	Magnitude
2			Lower	Best Estimate	Upper	Lower	Best Estimate	Upper	Lower
3	1								
4		1	1.00E-18	1.00E-14	1.00E-10	0.00E+00	0.00E+00	0.00E+00	0.00E+00
5		2	1.00E-16	1.00E-14	1.00E-08	1.00E+02	1.00E+03	1.00E+05	2.00E+02
6		3	1.00E-16	1.00E-14	1.00E-06	1.00E+03	1.00E+06	2.00E+06	5.00E+03
7		4	1.00E-09	1.00E-05	1.00E-03	0.00E+00	1.00E+00	1.00E+02	0.00E+00
8		5	1.00E-12	1.00E-05	1.00E-04	5.00E+02	1.00E+03	2.00E+03	5.00E+02
9		6	1.00E-06	1.00E-05	1.00E-04	1.45E+03	1.00E+06	1.50E+06	1.00E+05
10		7	1.00E-09	1.00E-02	1.00E-01	0.00E+00	0.00E+00	0.00E+00	0.00E+00
11		8	1.00E-05	1.00E-02	1.00E-02	1.00E+01	1.00E+03	2.00E+03	3.00E+02
12		9	1.00E-03	1.00E-02	1.00E-01	5.00E+02	1.00E+06	1.10E+06	1.00E+04
13	Upper Limit		1.01E-03		1.99E-01				
14	Lower Limit		1.00E-03		1.00E-01				
15	Best Estimate		1.00E-03		1.99E-01				
16	2								
17		1	1.00E-13	1.00E-10	1.00E-07	0.00E+00	0.00E+00	0.00E+00	1.00E+00
18		2	1.00E-10	1.00E-07	1.00E-07	2.00E+00	1.00E+03	3.00E+03	5.00E+01
19		3	1.00E-10	1.00E-08	1.00E-04	2.00E+02	1.00E+06	1.10E+06	1.00E+03
20		4	1.00E-10	1.00E-08	1.10E-07	0.00E+00	1.00E+00	1.00E+02	2.00E+01
21		5	1.00E-08	1.00E-07	1.00E-04	9.00E+02	1.00E+03	1.20E+03	0.00E+00
22		6	1.00E-04	1.00E-03	1.00E-02	1.00E+05	1.00E+06	1.10E+06	0.00E+00
23		7	1.00E-07	1.00E-06	1.00E-04	0.00E+00	0.00E+00	0.00E+00	1.00E+05
24		8	1.00E-13	1.00E-08	1.00E-05	5.00E+02	1.00E+03	7.00E+03	1.00E+03
25		9	1.00E-14	1.00E-09	1.00E-05	1.00E+05	1.00E+06	1.10E+06	9.99E+02
26	Upper Limit		1.00E-04		1.03E-02				
27	Lower Limit		1.00E-04		1.00E-02				
28	Best Estimate		1.00E-04		1.03E-02				

TABLE 2-8
INTERVAL ANALYSIS FOR SINGLE FAILURE MODES (CONT'D)

	J	K	L	M	N	O	P	Q
1	of	Damage	Human Risk		Economic Risk		Ranking according to	
2	Best Estimate	Upper	Lower	Upper	Lower	Upper	Human Risk	Economic Risk
3								
4	0.00E+00	0.00E+00	0.00E+00	0.00E+00	0.00E+00	0.00E+00	8	8
5	1.00E+03	2.00E+03	1.00E-14	1.00E-03	2.00E-14	2.00E-05	7	7
6	1.00E+06	1.50E+06	1.00E-13	2.00E+00	5.00E-13	1.50E+00	4	4
7	1.00E+00	1.00E+02	0.00E+00	1.00E-01	0.00E+00	1.00E-01	6	6
8	1.00E+03	5.00E+03	5.00E-10	2.00E-01	5.00E-10	5.00E-01	5	5
9	1.00E+06	1.90E+06	1.45E-03	1.50E+02	1.00E-01	1.90E+02	2	2
10	0.00E+00	0.00E+00	0.00E+00	0.00E+00	0.00E+00	0.00E+00	8	8
11	1.00E+03	7.00E+03	1.00E-04	2.00E+01	3.00E-03	7.00E+01	3	3
12	1.00E+06	1.80E+06	5.00E-01	1.10E+05	1.00E+01	1.80E+05	1	1
13								
14								
15			5.02E-01	1.10E+05	1.01E+01	1.80E+05	1	1
16								
17	1.00E+03	1.20E+03	0.00E+00	0.00E+00	1.00E-13	1.20E-04	8	7
18	1.00E+02	1.00E+03	2.00E-10	3.00E-04	5.00E-09	1.00E-04	6	8
19	5.00E+03	6.00E+03	2.00E-08	1.10E+02	1.00E-07	6.00E-01	2	4
20	1.00E+03	5.00E+03	0.00E+00	1.10E-05	2.00E-09	5.50E-04	7	6
21	1.00E+00	1.00E+02	9.00E-06	1.20E-01	0.00E+00	1.00E-02	4	5
22	0.00E+00	0.00E+00	1.00E+01	1.10E+04	0.00E+00	0.00E+00	1	9
23	1.00E+06	2.00E+06	0.00E+00	0.00E+00	1.00E-02	2.00E+02	8	1
24	1.00E+06	2.00E+06	5.00E-11	7.00E-02	1.00E-10	2.00E+01	5	2
25	1.00E+03	1.00E+05	1.00E-09	1.10E+01	9.99E-12	1.00E+00	3	3
26								
27								
28			1.00E+01	1.11E+04	1.00E-02	2.22E+02	2	2

The probability of system failure is expressed in the form of an interval with upper and lower limits. These limits are determined such that they cover all the uncertainties and correlation levels of the elements. The lower limit of the system's probability of failure is determined as the smaller value in the lower limit interval, as shown in the crossing of column C by rows 15 and 28 for systems 1 and 2, respectively, whereas the upper system limit is determined as the larger value in the upper limit interval as shown in the crossing of column E by rows 15 and 28 for systems 1 and 2, respectively. The ranking of elements and systems are determined using the logic of Eqs. 2-9 through 2-14, as shown in columns P and Q for human and economic risk, respectively.

Comparative Discussion. Figures 2-9 and 2-10 show a ranking comparison for system 1 elements according to single failure modes for human and economic risks, respectively. Figures 2-11 and 2-12 show a ranking comparison for system 2 elements according to single failure modes for human and economic risks, respectively. Figure 2-13 shows a comparison of system ranking for human and economic risks. The results shown in these figures demonstrate the effect of the method of analysis on the resulting ranking of elements and systems. The inclusion of uncertainty in the evaluation using the interval analysis can result in a change of the ranking of the elements or systems for inspection purposes.

Example 2: Multiple Failure Modes. In the case where some elements exhibit different failure modes, element ranking can be based on the complete ranking of all failure modes of the elements. In such cases, failure modes of all the elements are compiled in one evaluation process, and they are ranked based on either the single point estimate or the interval analysis. The evaluation is similar to the single failure modes as described in the previous section.

Single Point Estimate. Table 2-9 shows a system composed of six elements. Some of the elements exhibit single failure modes, others exhibit multiple failure modes. Each mode of failure has a single point estimate for the probability of failure. The magnitudes of casualty and damage are represented using single point estimates for each mode of failure. The human and economic risks for each mode of failure are assessed as the product of the failure probability by the magnitude of casualty and the magnitude of damage, respectively. The results are single point estimates. The failure modes are ranked accordingly.

Interval Analysis. Table 2-10 shows the same example in a similar format to Table 2-9. In this case, the occurrence probabilities of failure modes, and the magnitudes of casualty and damage are represented by intervals. The risk calculations are performed using the same approach for the single failure mode case. The resulting ranking of the risk intervals is shown in columns P and Q according to human and economic risks, respectively.

Comparative Discussion. Figures 2-14 and 2-15 show a ranking comparison for multiple failure modes based on single point estimate (Table 2-9) and interval analysis (Table 2-10) according to human and economic risk, respectively.

After ranking the failure modes for the components, a prioritized list of components can be developed. This component list can be established based on ranking the components according to the inspection priority of their failure modes. Therefore, a component with a higher ranked failure mode would have a higher ranking than other components.

Mathematical expressions for evaluating the entries in Table 2-7 through 2-10 are provided in Appendix B.

Group Ranking. In many cases, the analyst is more interested in ranking elements or systems in groups, where the elements or systems in each group have relatively equal estimated attributes, rather than ranking them individually. The group ranking can be based on the probability of failure, consequences, human, and/or economic risks. The group ranking results in establishing a priority list for inspection.

As an example of group ranking, the elements of system 1 introduced in Example 1 for single failure modes with single point estimates are ranked in groups according to the probability of failure, the magnitude of damage, and the economic risk as shown in Figs. 2-16, 2-17, and 2-18, respectively. For example, in Fig. 2-18 the top inclined arrow line separated and grouped elements 8, 9, and 6 in group A, which has an economic risk greater than 1.5. It should be noted that all points, which lay along one line, have the same economic risk. Figure 2-19 is a summary of different group ranking based on the probability of failure, the magnitude of damage, and the economic risk.

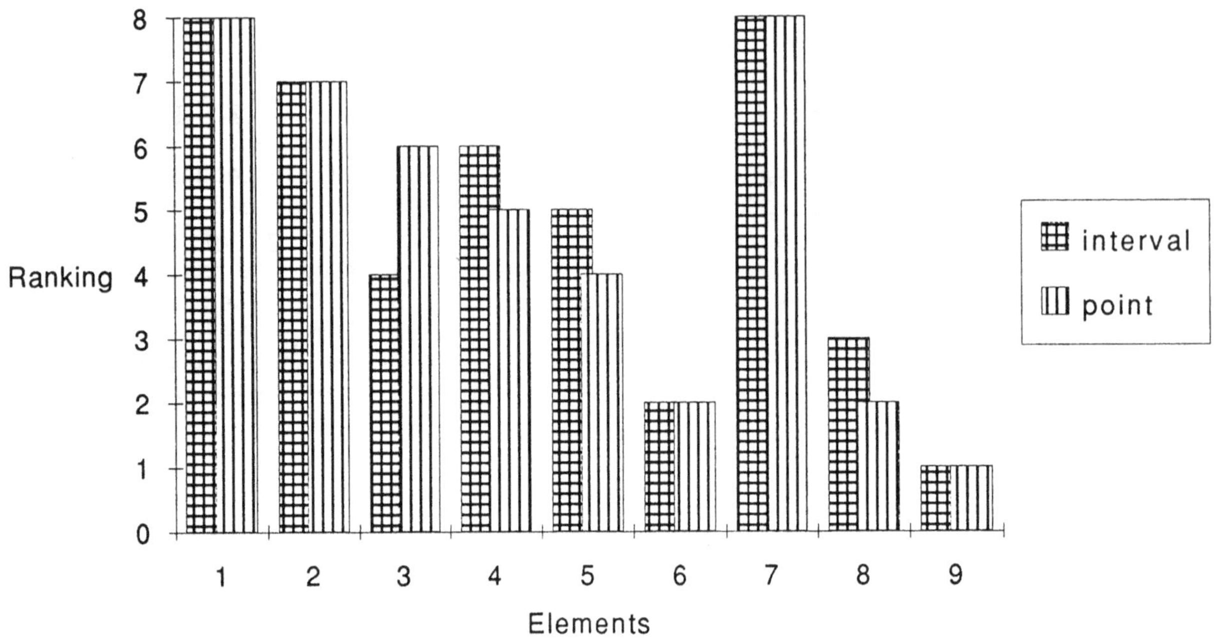

FIG. 2-9 RANKING OF SYSTEM 1 ELEMENTS FOR SINGLE FAILURE MODES BASED ON HUMAN RISK

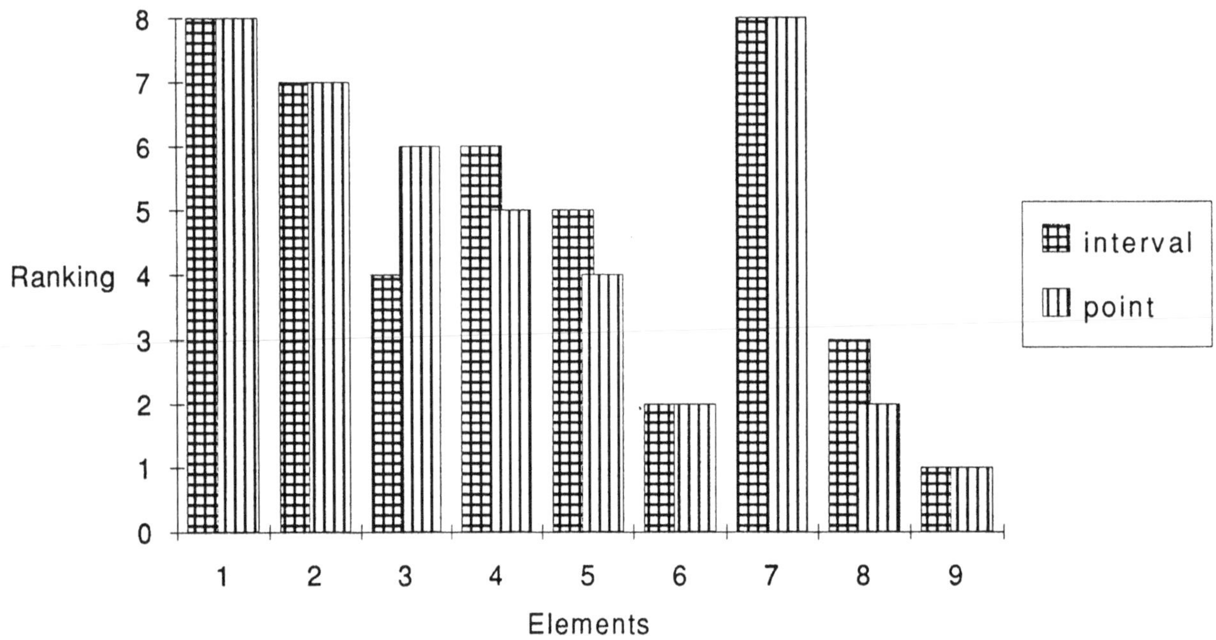

FIG. 2-10 RANKING OF SYSTEM 1 ELEMENTS FOR SINGLE FAILURE MODES BASED ON ECONOMIC RISK

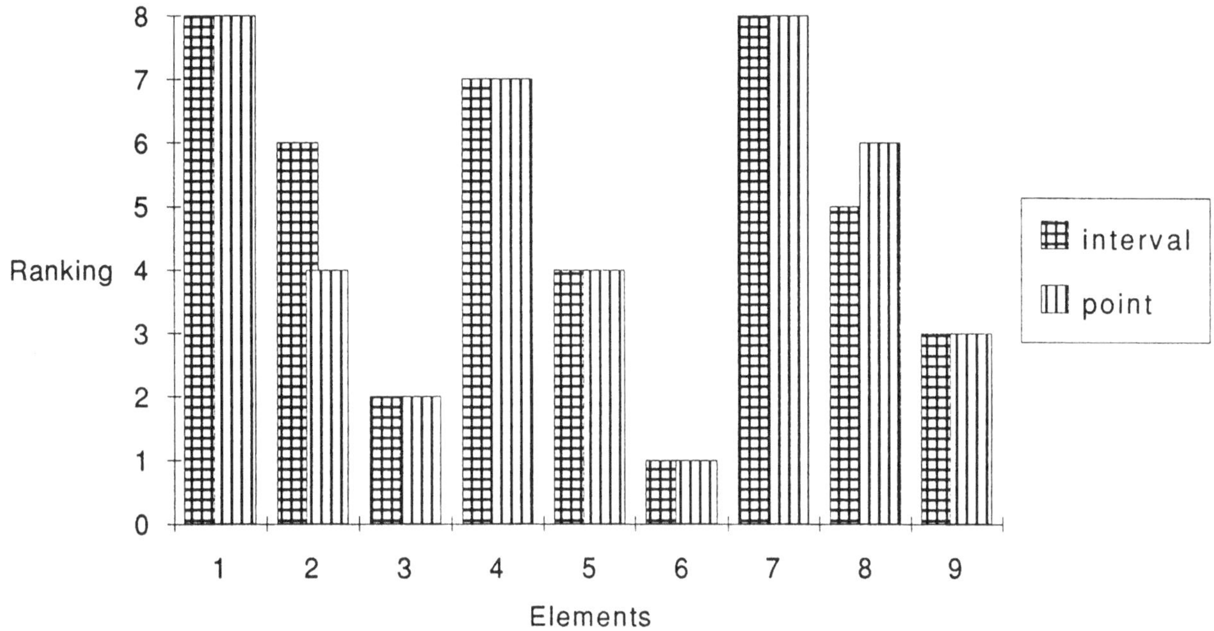

FIG. 2-11 RANKING OF SYSTEM 2 ELEMENTS FOR SINGLE FAILURE MODES BASED ON HUMAN RISK

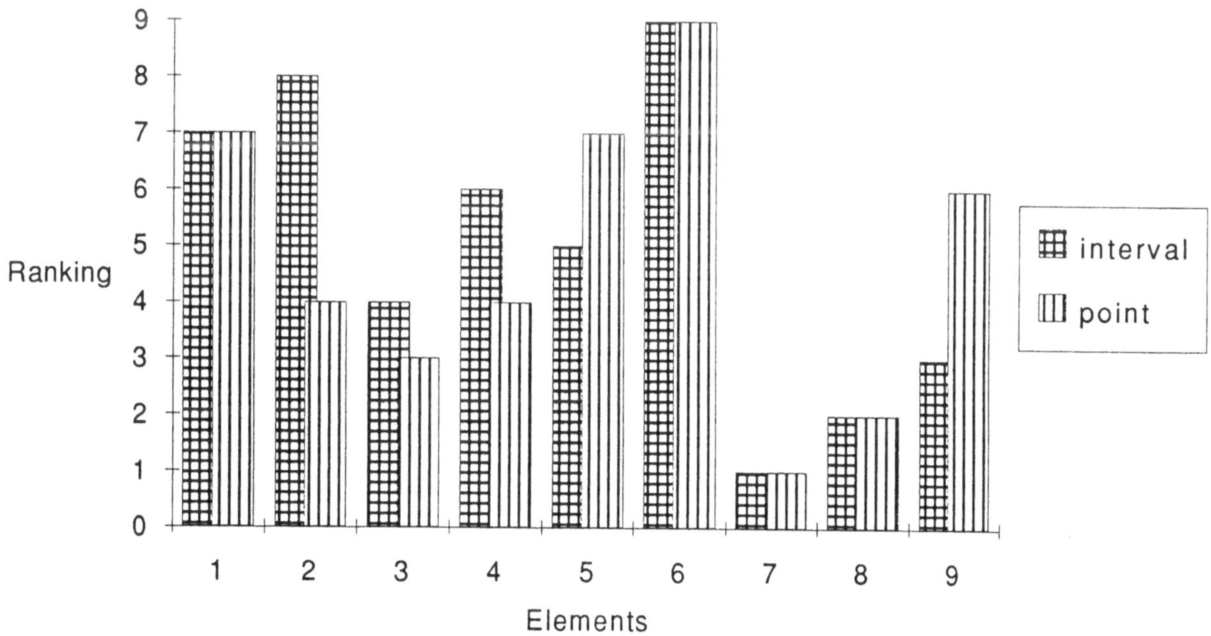

FIG. 2-12 RANKING OF SYSTEM 2 ELEMENTS FOR SINGLE FAILURE MODES BASED ON ECONOMIC RISK

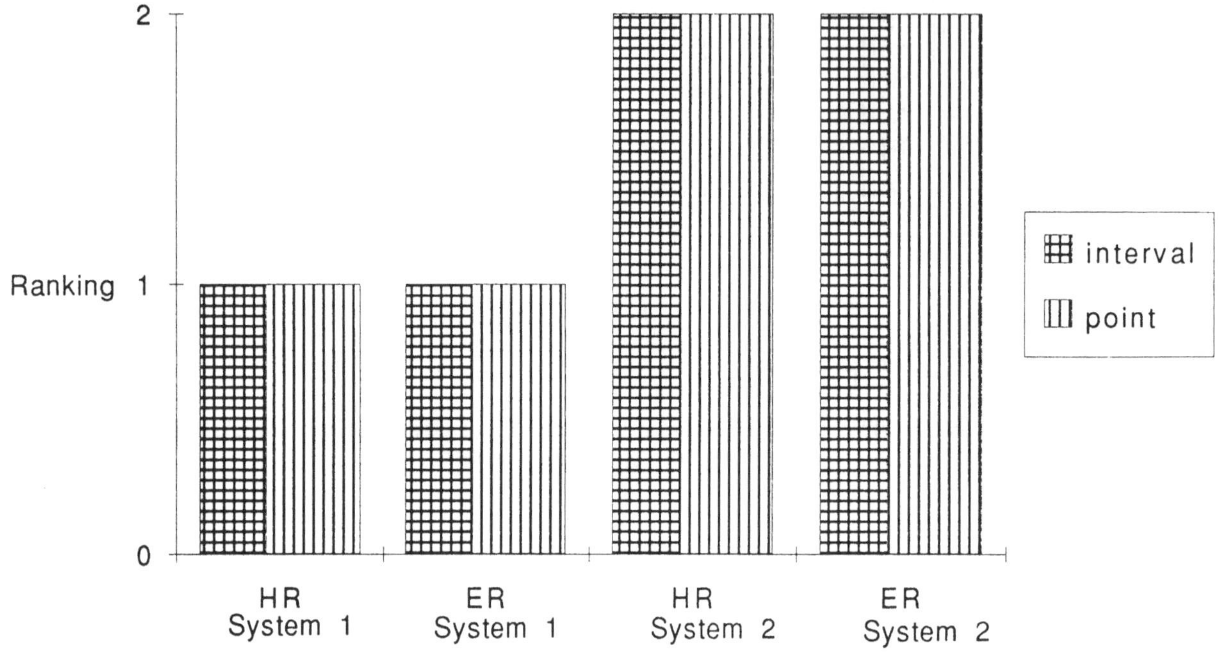

FIG. 2-13 SYSTEM RANKING BASED ON HUMAN RISK (HR) AND ECONOMIC RISK (ER)

TABLE 2-9
POINT ESTIMATE METHOD FOR MULTIPLE FAILURE MODES

	A	B	C	D	E	F	G	H	I
1	Element	Failure	Probability	Magnitude of				Ranking according to	
2		Mode	of Faliure	Casualties	Damage	Human Risk	Economic Risk	Human Risk	Economic Risk
3	1	Mode 1	1.00E-13	0.00E+00	2.00E+02	0.00E+00	2.00E-11	8	10
4		Mode 2	1.00E-04	0.00E+00	1.00E+02	0.00E+00	1.00E-02	8	1
5	2	Mode 1	1.00E-07	1.00E+05	1.00E+04	1.00E-02	1.00E-03	1	2
6	3	Mode 1	1.00E-10	1.00E+03	3.50E+03	1.00E-07	3.50E-07	7	7
7		Mode 2	1.00E-08	3.00E+02	2.00E+02	3.00E-06	2.00E-06	3	5
8	4	Mode 1	1.00E-08	3.40E+01	7.00E+02	3.40E-07	7.00E-06	6	4
9		Mode 2	1.00E-12	0.00E+00	5.00E+02	0.00E+00	5.00E-10	8	9
10		Mode 3	1.00E-08	5.00E+01	9.00E+01	5.00E-07	9.00E-07	4	6
11	5	Mode 1	1.00E-05	7.50E+02	4.50E+01	7.50E-03	4.50E-04	2	3
12	6	Mode 1	1.00E-10	4.00E+03	4.30E+02	4.00E-07	4.30E-08	5	8

TABLE 2-10
INTERVAL ANALYSIS FOR MULTIPLE FAILURE MODES

	A	B	C	D	E	F	G	H
1	Element	Failure Mode	Probability of Failure			Magnitude of Casualties		
2								
3			Min	Best Estimate	Max	Min	Best Estimate	Max
4	1	Mode 1	1.00E-16	1.00E-13	1.00E-08	0.00E+00	0.00E+00	3.00E+01
5		Mode 2	1.00E-06	1.00E-04	1.00E-03	0.00E+00	0.00E+00	0.00E+00
6	2	Mode 1	1.00E-09	1.00E-07	1.00E-05	3.00E+03	1.00E+05	5.00E+05
7	3	Mode 1	1.00E-12	1.00E-10	1.00E-07	3.00E+01	1.00E+03	5.00E+04
8		Mode 2	1.00E-10	1.00E-08	1.00E-06	2.00E+02	3.00E+02	3.50E+02
9	4	Mode 1	1.00E-09	1.00E-08	1.00E-05	2.00E+01	3.40E+01	5.00E+02
10		Mode 2	1.00E-12	1.00E-12	1.00E-09	0.00E+00	0.00E+00	1.20E+01
11		Mode 3	1.00E-12	1.00E-08	1.00E-06	4.00E+01	5.00E+01	6.00E+01
12	5	Mode 1	1.00E-10	1.00E-05	1.00E-04	5.00E+02	7.50E+02	7.00E+04
13	6	Mode 1	1.00E-13	1.00E-10	1.00E-09	2.00E+03	4.00E+03	8.00E+04

	I	J	K	L	M	N	O	P	Q
1	Magnitude of							Ranking according to	
2	Damage			Human Risk		Economic Risk		Human Risk	Economic Risk
3	Min	Best Estimate	Max	Min	Max	Min	Max		
4	7.00E+01	2.00E+02	3.00E+03	0.00E+00	3.00E-07	7.00E-15	3.00E-05	8	9
5	9.00E+01	1.00E+02	1.50E+02	0.00E+00	0.00E+00	9.00E-05	1.50E-01	10	2
6	2.00E+03	1.00E+04	1.00E+05	3.00E-06	5.00E+00	2.00E-06	1.00E+00	2	1
7	7.00E+02	3.50E+03	4.00E+03	3.00E-11	5.00E-03	7.00E-10	4.00E-04	4	7
8	2.00E+01	2.00E+02	5.00E+02	2.00E-08	3.50E-04	2.00E-09	5.00E-04	5	6
9	5.00E+02	7.00E+02	7.00E+03	2.00E-08	5.00E-03	5.00E-07	7.00E-02	3	3
10	4.00E+02	5.00E+02	5.50E+02	0.00E+00	1.20E-08	4.00E-10	5.50E-07	9	10
11	1.00E+01	9.00E+01	1.00E+02	4.00E-11	6.00E-05	1.00E-11	1.00E-04	7	8
12	3.00E+01	4.50E+01	3.00E+02	5.00E-08	7.00E+00	3.00E-09	3.00E-02	1	4
13	4.00E+02	4.30E+02	1.00E+06	2.00E-10	8.00E-05	4.00E-11	1.00E-03	6	5

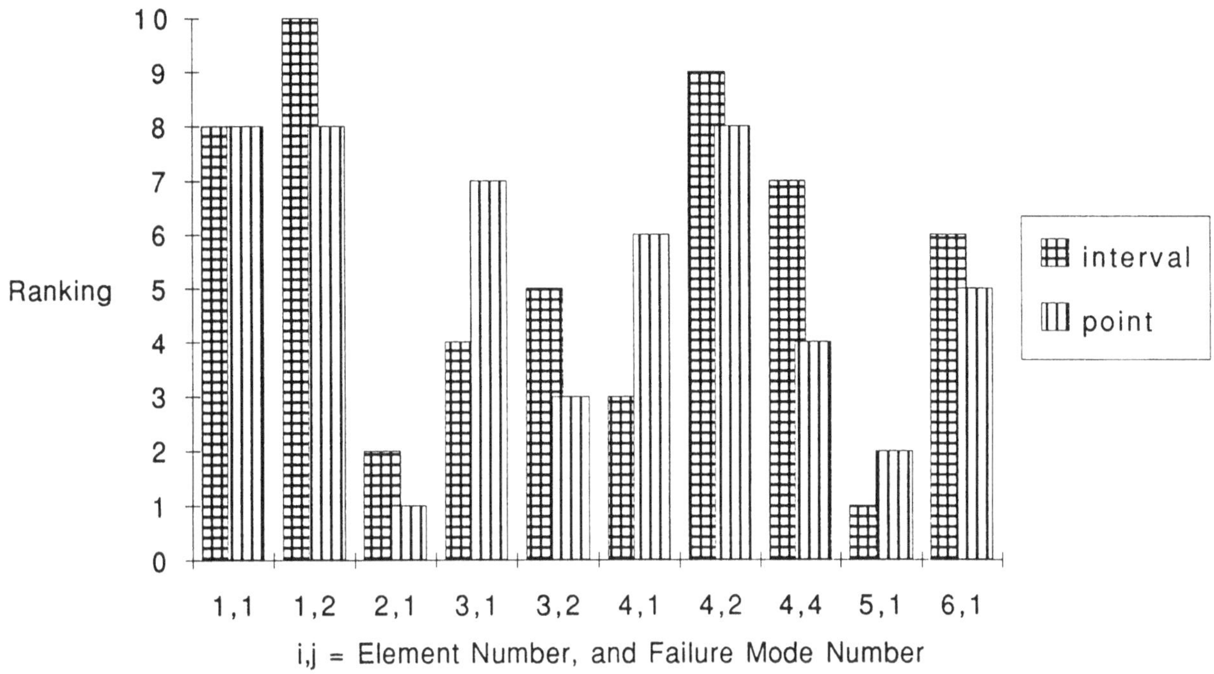

FIG. 2-14 RANKING OF MULTIPLE FAILURE MODES BASED ON HUMAN RISK

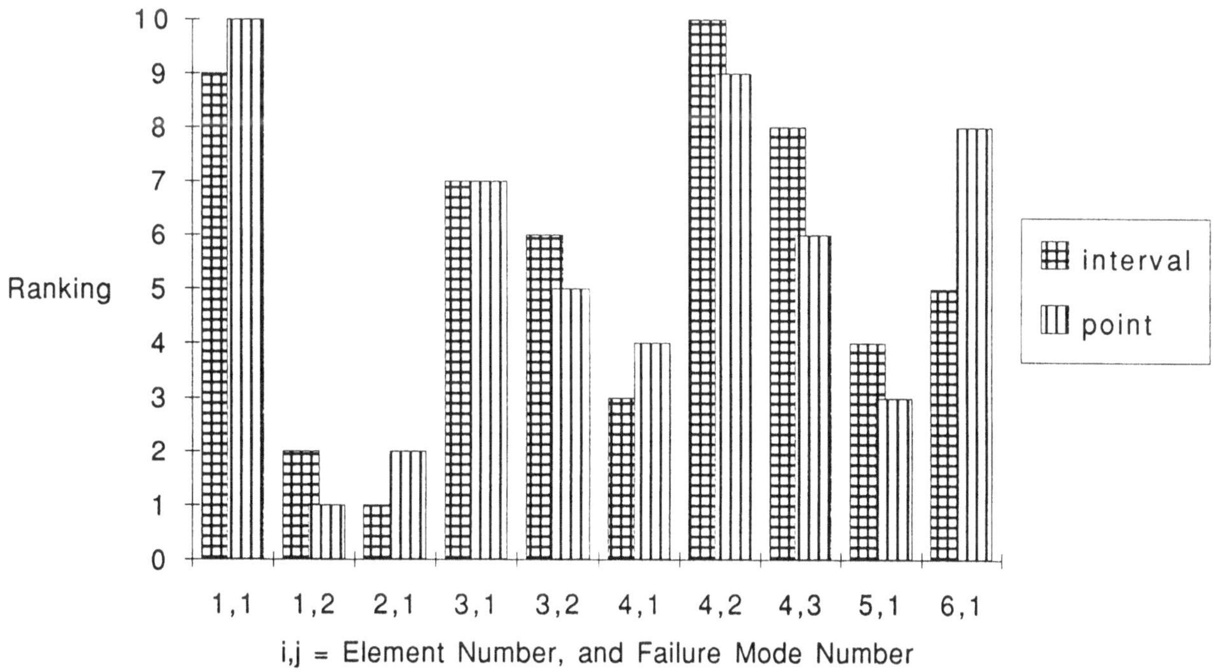

FIG. 2-15 RANKING OF MULTIPLE FAILURE MODES BASED ON ECONOMIC RISK

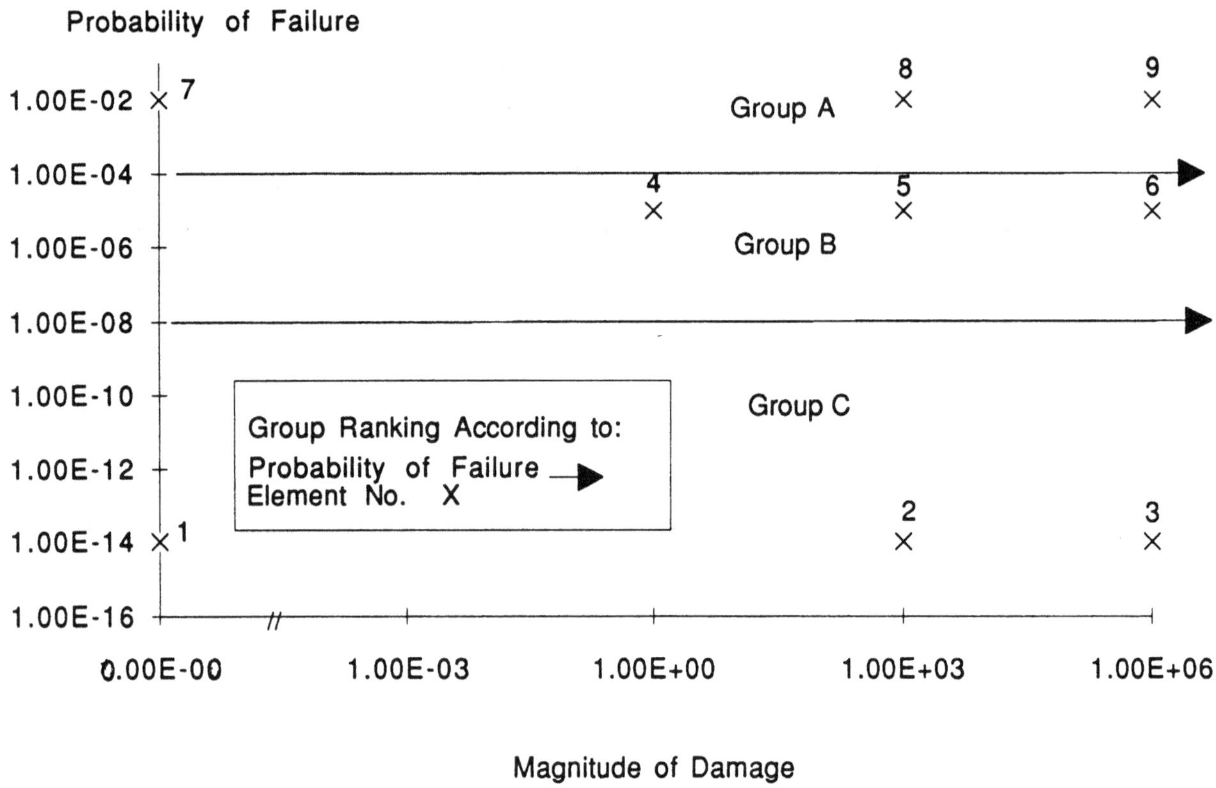

FIG. 2-16 GROUP RANKING ACCORDING TO PROBABILITY OF FAILURE

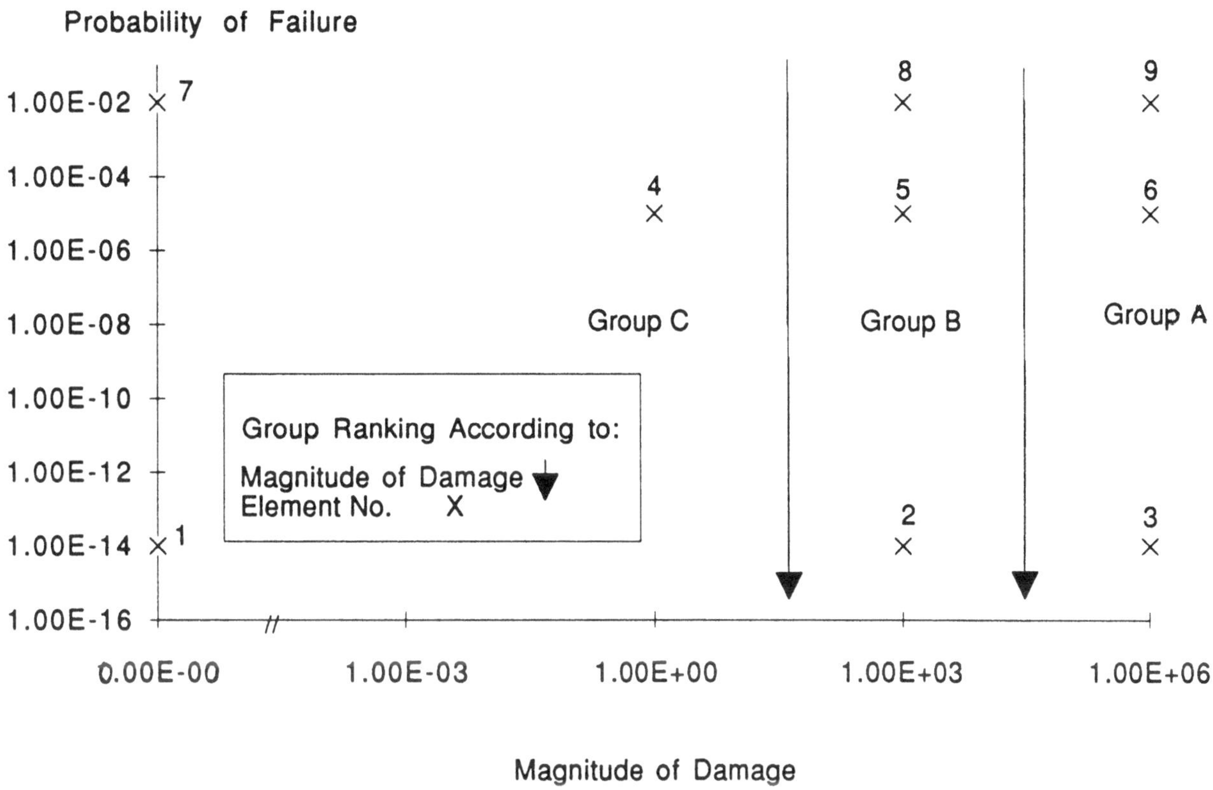

FIG. 2-17 GROUP RANKING ACCORDING TO MAGNITUDE OF DAMAGE

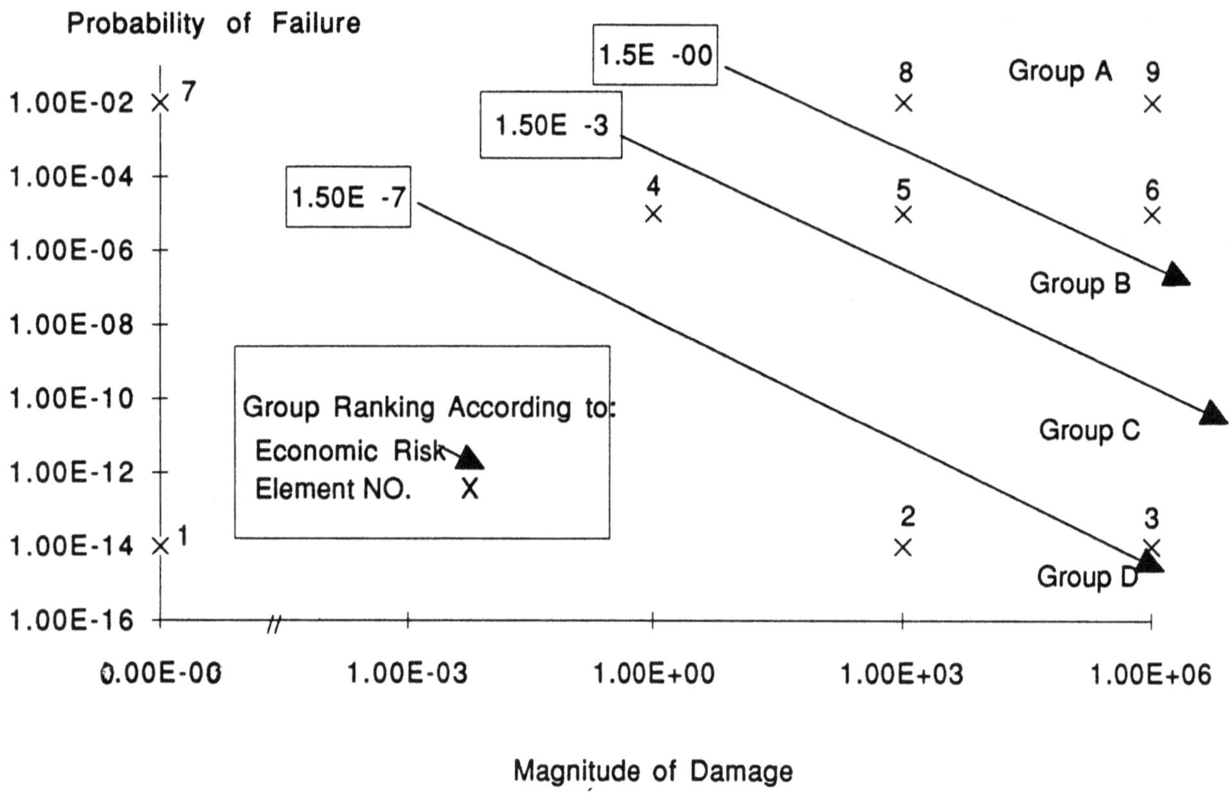

FIG. 2-18 GROUP RANKING ACCORDING TO ECONOMIC RISK

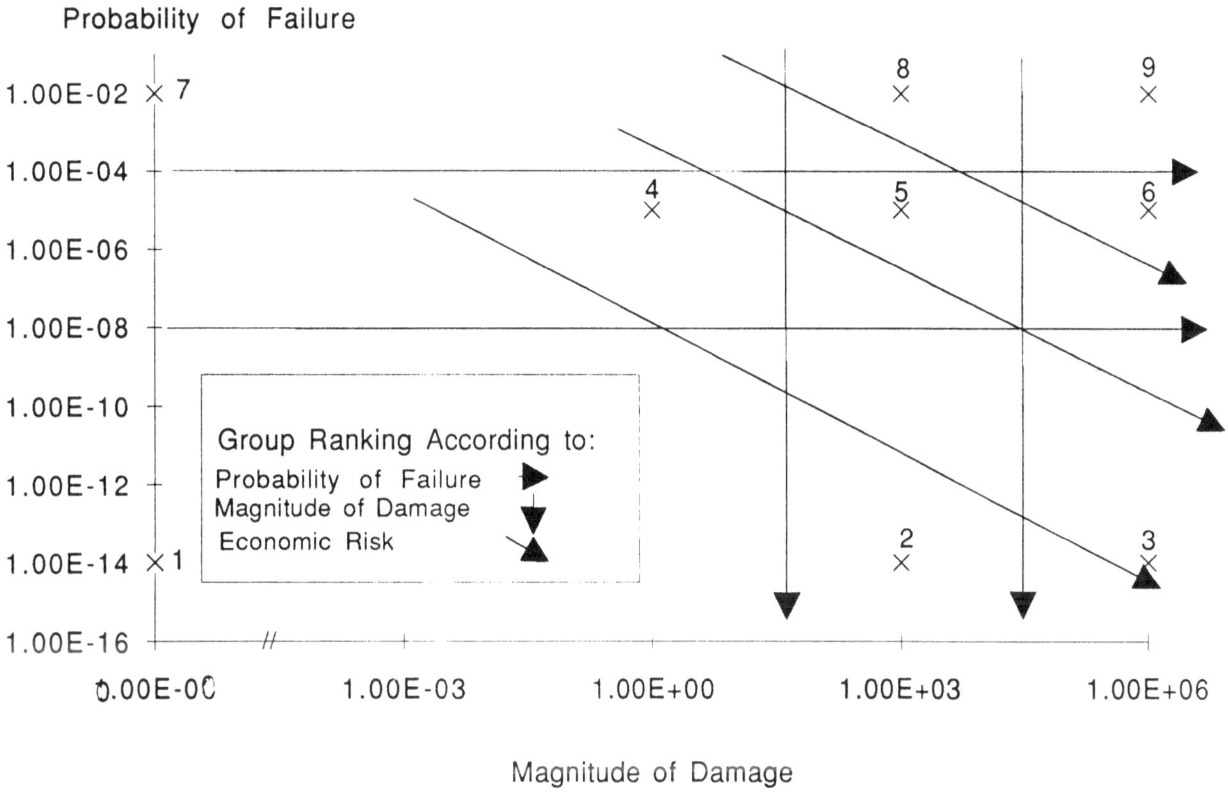

FIG. 2-19 GROUP RANKING SUMMARY

2.6 ☐ INSPECTION PROGRAM DEVELOPMENT

2.6.1 ☐ Overview

Once components or structural elements are ranked or categorized in the first part of the quantitative risk analysis, the next step in the overall process is to develop an inspection program for each group of components or elements. This process, which is shown in Fig. 2-20, can also be used to establish an inspection program for an individual component or system, as necessary. The recommended process is divided into a series of three basic steps.

(1) Choose Candidate Inspection Strategies That Define the Frequency, Method, and Sampling Procedure for Inspection. The method of inspection includes the procedure, tool, and level of personnel qualification to perform the inspection. The inspection strategy may also take advantage of monitoring systems (e.g., for thermal effects) and maintenance testing programs. Critical uncertainties associated with this step are the potential for damage states to exist in the component or element, the potential for inspection damage (which also includes the potential for danger to the inspector), and the reliability of the inspection method, including the potential for false calls.

For example, if a situation exists for a component where the chance for potential damage is remote, the inspection requires the movement of heavy equipment and causes the inspector to be exposed to a hazardous environment, and where the reliability of the best available method for finding the damage state is 50/50 and usually results in false calls, common sense indicates that "no inspection" is a viable alternative until less threatening and more effective inspection techniques are developed.

(2) Choose an Inspection Strategy and Perform Inspection. From the candidate inspection strategies, defined in Step 1, the effect of each of these strategies on the failure probability of the component or element of interest is estimated. The key uncertainties to be considered in this estimate are the inspection reliability, the chance that certain degradation mechanisms are occurring, the potential for certain levels of loads to occur, and the potential failure mode of the component or element. Structural reliability and risk assessment (SRRA) models can be used prior to and following the inspection in order to evaluate the sensitivity of these uncertainties relative to the impact of the candidate inspection strategies on the failure probabilities. An inspection strategy is chosen based on these results, and the inspection is performed.

(3) Choose Appropriate Action and Update State-of-Knowledge. Following the performance of the inspection, another critical decision is faced. That is, should the component or element be repaired or replaced if significant findings occur, or should nothing be done except to redefine the inspection strategy back in Part 1? If a repair or replacement is required, another decision that is faced is whether to take the action now or later. This depends on whether this action will indeed be successful for the intended period of operation, and whether the potential exists for new damage to be introduced. The SRRA process can be used once again to determine the effects of findings and potential corrective actions on the failure probabilities. In any case, all of the results related to the inspection should be used to update the FMECA information on a periodic basis to rerank the components or elements based on risk and redefine the inspection strategy in Part 1, providing a "living process" as long as the component or element is in service.

The subsections given below discuss more details on each of these three parts of the inspection program development through the use of decision risk-analysis logic trees. The logic appears complex at first, but once one walks through the process, an appreciation is obtained for the critical questions that must be addressed when establishing an inspection program. A brief overview and simple tutorial example of decision analysis are provided next to help better understand the logic process.

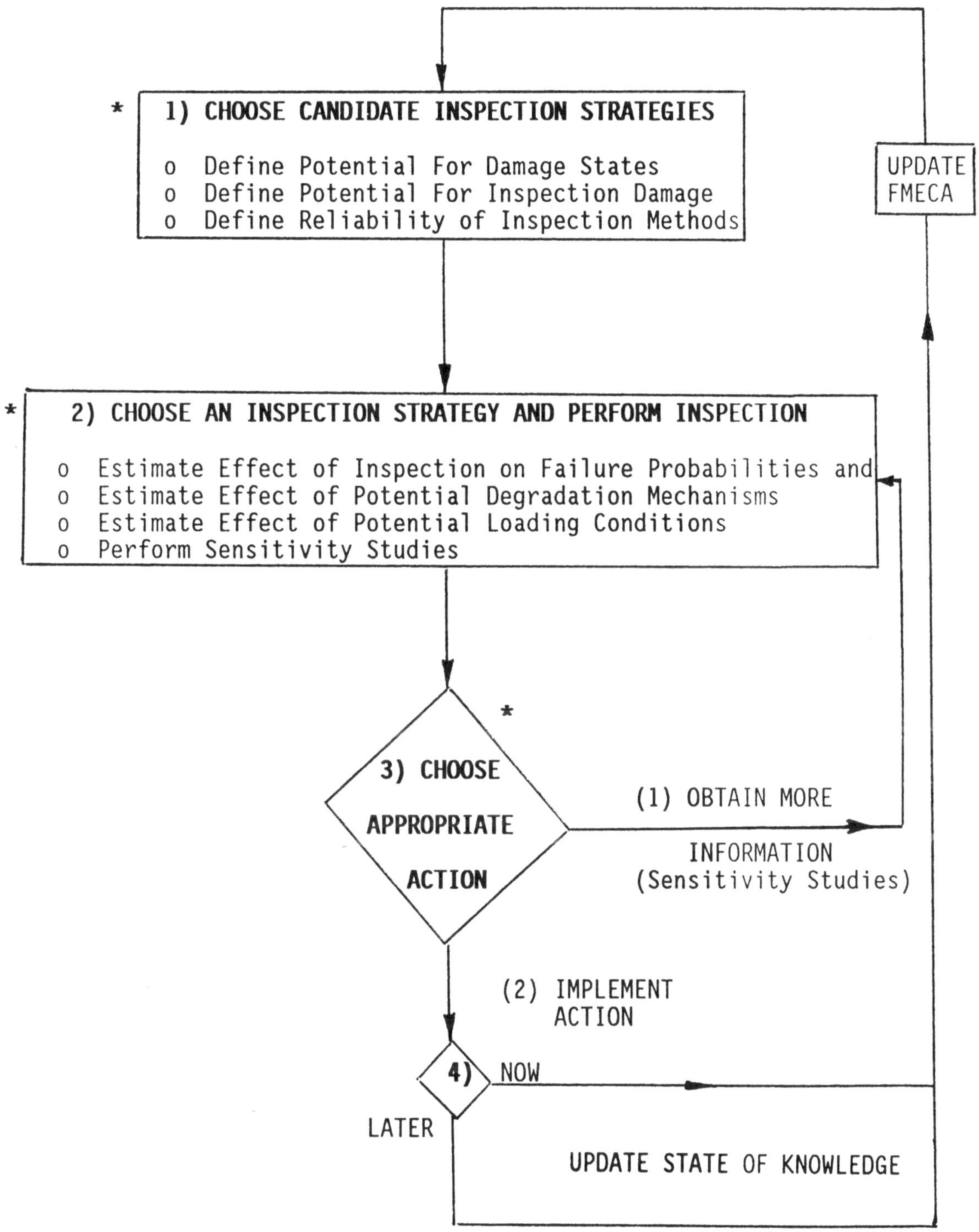

* Decision Risk Analysis

FIG. 2-20 INSPECTION PROGRAM DEVELOPMENT

The Decision Analysis Approach.[a] Decision analysis is a prescriptive methodology for improving the chances of making rational choices in the face of uncertainty and risk. With its roots in Bayesian probability theory, decision analysis also provides a common language for talking about "chances," "preferences," "risks," and the like – a feature that should be of interest to those seeking a holistic, cross-disciplinary approach. The development of inspection programs possesses all the attributes of a good decision analysis problem – namely, complexity, uncertainty, and risk. Doubtless, much of the uncertainty can be resolved or reduced through further research results, but one usually does not have the luxury of waiting for scientific breakthrough.

The argument for including decision analysis in the risk-based inspection process rests on the method's demonstrated ability to deal with a generic problem; to wit, much of the objective data required to make really strategic decisions is often unavailable when it is needed. Decision analysis has proven to be a powerful technique for making such "choices under uncertainty" in a variety of problem domains. [Brown, et al. (1974) provides an accessible exposition on decision analysis and its applications.]

Decision analysis focuses on finding a decision that is "rational" in the sense of incorporating a decision-making unit's state-of-knowledge about uncertain quantities and its attitude toward risk. The process by which this goal is reached can be described in terms of a "decision analysis cycle" consisting of four sequential stages: (1) problem-structuring, (2) deterministic (e.g., systems engineering) modeling, (3) probabilistic analysis, and (4) evaluation.

The first stage identifies the problem in terms of alternatives, decision criteria, uncertainties, and the sequence of decisions to be made. A tool called the "influence diagram" is used to fully specify the algebraic representation of the problem (i.e., the model). A decision tree is constructed which concisely displays the technically feasible alternatives and the range of possible scenarios. The second stage involves construction of a model which deterministically calculates the value of the decision criteria for any chosen scenario (i.e., path through the tree). The subsequent probabilistic stage involves the following:

(a) encoding uncertainty as a discretized probability distribution at each of the tree's "chance nodes," primarily from elicitation of expert opinion;

(b) identifying major uncertainties via probabilistic sensitivity analysis;

(c) solving the decision tree to obtain probability distributions and expected (or, if necessary, risk-adjusted) values for the decision criterion.

The final or evaluation stage helps the decision-maker decide whether action should be taken now or postponed in favor of further information-gathering and analysis. Tools here include the "value of information" and "value of control" techniques borrowed from statistical decision theory.

Figure 2-21 provides a simple tutorial example in which a plant is considering adopting a "new" inspection method for detecting repairable fatigue damage in a section of high-pressure steam piping. A strategy of "no inspection" is evaluated so that a potential relaxation from the current method is also considered. The new method has a higher detection probability than the "current" method, but also a higher implementation cost for inspection. No inspection obviously has a nil detection probability and no implementation cost.

The decision tree illustrates the sequence of decisions and uncertainties involved in the choice between current methods, new methods, and no inspection. Following any particular path through the tree leads to a single value of the decision criterion (total cost). The probabilities attached to the branches at each chance node represent the

---

[a]Adapted from paper by Perdue (1988).

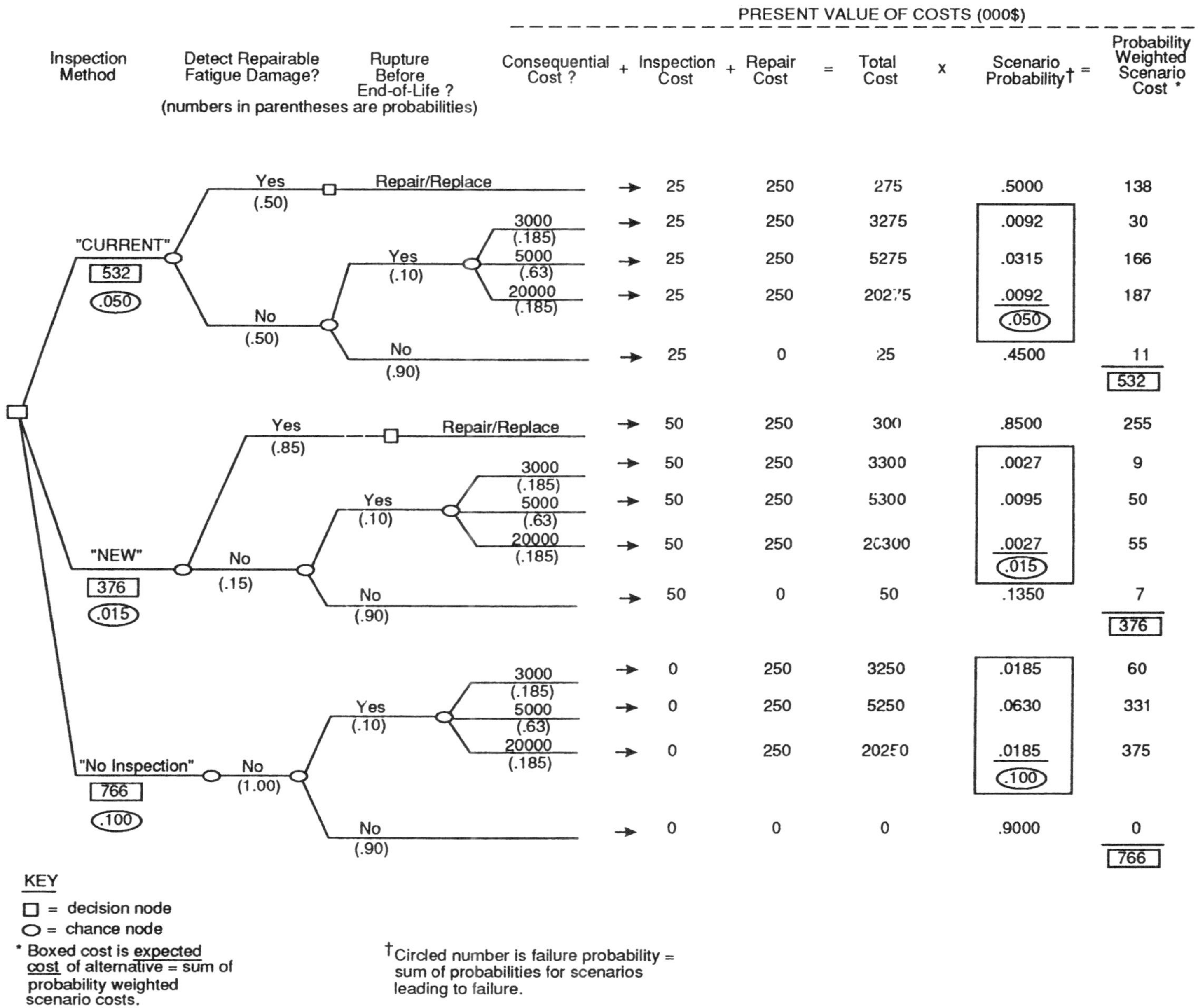

FIG. 2-21 ILLUSTRATIVE DECISION TREE FOR CHOOSING AN INSPECTION STRATEGY

likelihood of following that path. By starting at the top of the tree and following a process of taking expected values at chance nodes and optimizing (i.e., choosing the highest expected value) at decision nodes, the tree is usually "averaged out and folded back" to yield an expected total cost for each method. For the sake of example, the numerical calculations are shown to the right of the tree along with the path scenario probabilities, many of which can be used to evaluate an acceptable failure probability level by the user. In this case, the new method is seen to have the lowest expected cost ($376K) versus the current method ($532K). The strategy of "no inspection" yields the highest expected cost ($766K) and is dropped from further consideration. This strategy also yields the highest failure probabilities. Examination of the tree reveals that the probability of a "Rupture Before End of Life" is high enough that the new method avoids sufficient "consequential costs" to offset its higher "inspection cost." Further analysis would reveal the decision lumped on the uncertainty at the "Rupture Before End of Life" (or some other uncertainty) and could even provide estimates of the dollar amount that the plant should invest in information-gathering activities directed toward resolving the critical uncertainties. Finally, the decision-maker's "risk aversion" (e.g., to the possibility of following the path to the $20 million consequential cost) could be formally incorporated into the final decision.

The failure probabilities, even for the new inspection method, may be considered to be unacceptable by the user. Additional strategies, possibly considering more frequent inspections or including monitoring systems, may be developed in an attempt to yield acceptable failure probabilities. If acceptable failure probabilities cannot be achieved by any inspection strategy, the user now faces repair or replace decisions before carrying the inspection process any further. For many applications, SRRA tools are needed to evaluate the failure probabilities, particularly when these values are below those that can be reasonably obtained from expert opinion.

In summary, prudent management of inspection programs requires that the technical information resident in SRRA and other engineering tools be integrated with financial, regulatory, and other information into a comprehensive framework for evaluating alternatives and communicating results to management. Decision analysis can provide that framework.

2.6.2 ☐ Choose Candidate Inspection Strategies (See Fig. 2-22.)

Inspections are performed to identify and track the degradation of systems and components. This knowledge allows repair or remediation before the degradation progresses to failure. Depending on the situation, failure may be a flaw exceeding a defined critical flaw size, a leak of a certain size, or catastrophic rupture. Also, one must consider how the failure progresses, e.g., leak before break or break without leak. Since continuous monitoring of all potential failure sites is generally not feasible, strategies have been developed to infer this information from less complete information. This involves sampling, both among potential failure sites and in time. Thus, inspection strategies are actually strategies for sampling degradation information, which are optimized to provide a high-detection probability at an acceptable cost.

Figure 2-23 depicts the role of inspection over the service life of systems and components, and indicates how the inspection strategy can change over this time period. The linear path across the lower part of the diagram describes the strategy as projected at the beginning of service life. The original design and construction is expected to assure reliable operation to the end of life, and periodic inspections are performed only to gain added assurance that there is no significant degradation of structural integrity. The other paths through the diagram indicate actions and revised inspection

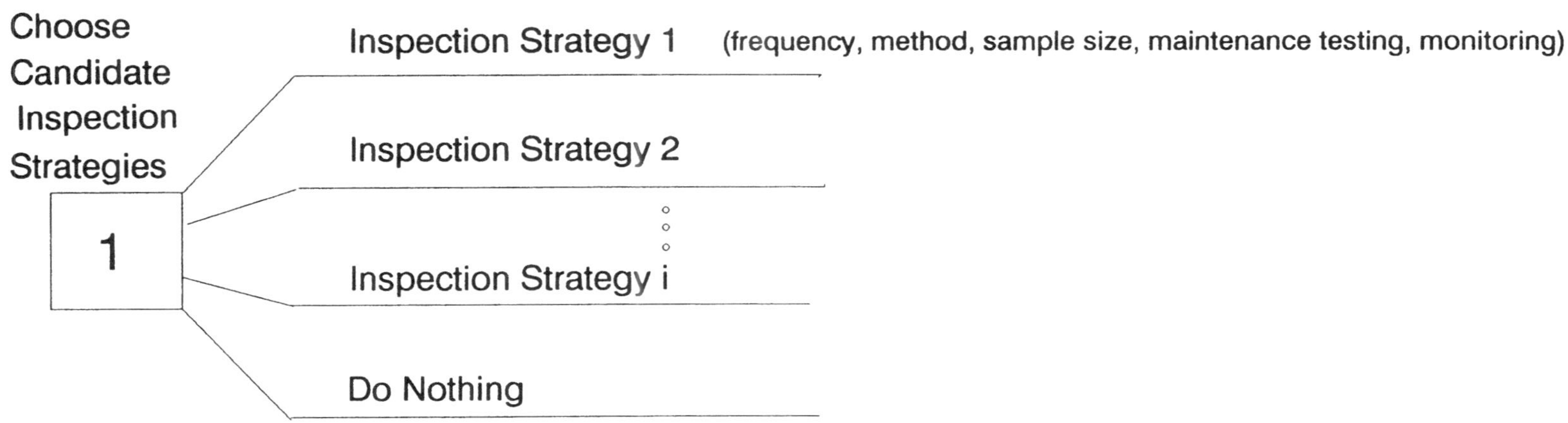

Driven by: Codes, Events, Time

Selection Criteria: Expected Risk and Costs

FIG. 2-22 SKELETON DECISION TREE FOR CHOOSING CANDIDATE INSPECTION STRATEGIES (DECISION NODE 1 ONLY)

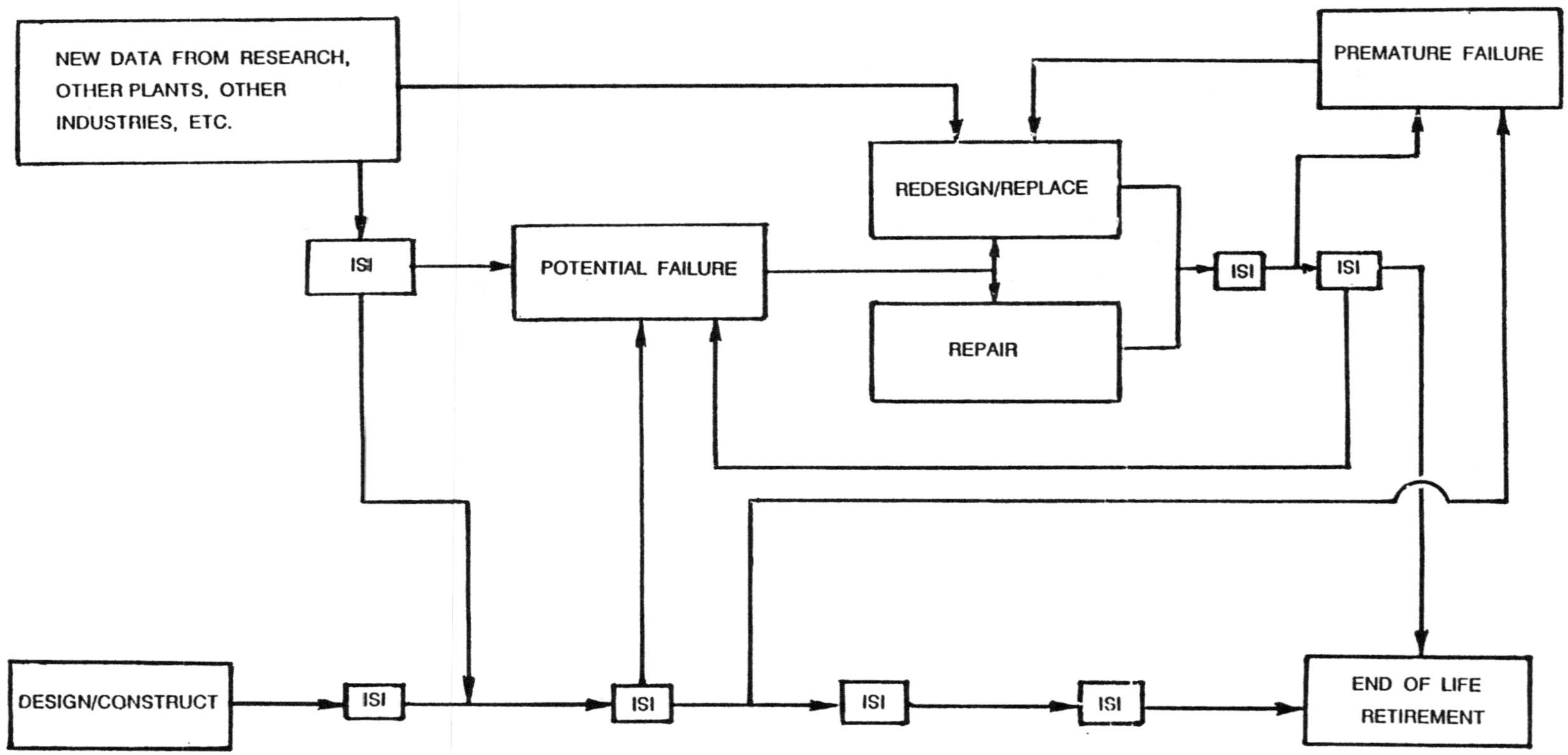

FIG. 2-23 ROLE OF INSERVICE INSPECTION (ISI) OVER SERVICE LIFE OF SYSTEMS AND COMPONENTS

strategies that result if new information reduces confidence in the integrity of the system or component. Such new information can come from such sources as premature failures, inspection findings that suggest potential failures, an awareness of more severe operating conditions than projected during design, identification of new degradation mechanisms, and results of research studies.

The probability of detecting degradation before failure depends on three primary factors. These are the time between inspections, the probability that the inspected sample contains the degrading component, and the probability that the inspection method detects the degradation in progress. Furthermore, all failures are not equal from a risk standpoint, so there is an incentive to bias the sampling process toward components whose failure is associated with a high risk. Finally, the consequences of some failures may be so high that they cannot be accommodated (e.g., financial ruin of a company or an industry), leading to a different biasing of the sampling process.

The time between inspections must be shorter than the time required for degradation to failure. Laboratory and field data, in combination with design and predictive analyses, can be used to determine intervals allowing multiple inspection samplings before failure is reasonably expected. In some cases, however, failure may occur rapidly after the onset of an adverse environmental condition (e.g., leaks in piping systems experiencing stress corrosion cracking). In such cases, more frequent environmental monitoring may be required between the regularly scheduled inspections. It may also be advantageous to vary the interval between inspections if failure rates are dependent on component aging.

In general, an inspection addresses a fraction of the components whose condition needs to be monitored. It is assumed that if the conditions of the inspected components are acceptable, this result is representative of the remainder of the components. This assumption is based on the expectation that components subject to similar physical and environmental effects will exhibit similar rates and patterns of degradation. Similarly, if the condition of any of the inspected components is unacceptable, the sample of inspected components should be enlarged, with complete inspection required if successive failures are exhibited. Factors involved in determining the fraction of components to be inspected include the variability in the environmental conditions and stress values, the variability in the degradation rates found among components subjected to similar stress conditions, the time between inspections, and the economic impact of the inspection process on operations (e.g., a process shutdown to allow inspection may by itself have a large economic impact, with little cost difference between a partial and a complete inspection once the shutdown occurs).

The sensitivity of the inspection process is of central importance to determination of the inspection strategy. A highly sensitive inspection process provides a high probability that incipient failures are detected. In addition, it allows the early detection and monitoring of degradation, providing confirmation that inspection intervals are adequately short, and that degradation is proceeding at expected rates. While a more sensitive inspection process does not provide assurance that a particular inspection sample size is adequate, the improved understanding of component status that it provides helps improve the effectiveness of any given sample size.

A less sensitive production-oriented inspection process, which can only detect degradation to the point of incipient failure, requires more frequent inspections. This is because the inspection must occur during the shorter period of time between the detection threshold and component failure. It may also require a larger sample size, to enhance the probability of early detection, as degradation approaches the threshold of detection. In this regard, a more sophisticated inspection at higher cost can increase the period of time by detecting degradation at a lower threshold.

Several factors are involved in selecting the sample of components to be inspected. Accessibility and potential hazards to the inspectors are two. Where interpretation of the inspection results may be ambiguous, there is reason to inspect the same sample during subsequent inspections so that successive results can be compared to detect changes. Where stressor conditions vary between components, there is reason to vary the inspected sample to attain complete coverage over several inspection cycles.

The risk of an ultimate undesirable consequence (e.g., structural collapse, reactor core meltdown, catastrophic environmental damage, large toxic release) to which component failures may contribute is a measure which can be used in selecting the sample of components to be inspected. Thus, if a large fraction of the total risk involves failures of a few components, there is reason to specifically include these components in the sample to be inspected so that as much specific information about the condition of these risk-important components is generated.

This common sense approach is straightforward to apply when risk is calculated using the FMECA approach described in Section 2.5. In situations where combinations of failures cause the undesirable consequences, a more sophisticated approach is required. One example is the Fussel-Vesely (F-V) Risk Importance Measure used by risk analysts, as discussed by Vesely, et al. (1983). For each component, its F-V importance is simply the fraction of the total risk to which its failures contribute. The F-V importance is calculated using the results of a probabilistic risk assessment (PRA). The PRA identifies specific combinations of component failures that lead to an undesired consequence (*cut sets*), and calculates the probability of each cut set occurring per year (cut set frequency). The F-V importance of any component is calculated by identifying all of the cut sets that involve the component, totalling the frequencies of these cut sets, and dividing by the sum of the frequencies of all cut sets.

Depending on the PRA methodology, each cut set may have a different consequence. This is the case in the analysis of reactor core damage accidents, when the radiation release due to core damage depends on the timing and sequencing of events. In this case, each cut set frequency is multiplied by its associated consequence, yielding a risk value for the cut set (e.g., person-rem/year). The F-V importance of a component is then the fraction of the total risk contributed by cut sets involving the component.

When selecting a sample of components for inspection, then, the selection process would be biased to include more of the risk-important components than would be the case for random selection. Several methods for this are possible, including proportional weighting, or weighting according to the order of magnitude of the risk involved.

Another consideration in defining potential inspection strategies is whether an inspection plan is being developed for structural components in one facility or for structural components in a set of similar facilities. In this case, the elements of the inspection plan are likely to differ as discussed below. Discussion of inspection of structural components of an individual aircraft versus a fleet of aircraft is used first to exemplify these special considerations.

Individual Versus Set of Components. In the military aircraft industry, the life of individual aircraft is determined by tracking the crack growth behavior of a number of critical points at which inspections are conducted on a periodic basis. Each aircraft has the same set of critical points; however, the crack growth rate, and hence the age of each aircraft, are different. This is due to the difference in how each aircraft is flown.

The aircraft design is based upon a baseline load spectrum. The aircraft must exhibit a given number of hours of life under the influence of this baseline spectrum. The actual flight usage of the aircraft is typically quite different from this baseline, depending upon the types of missions flown and the characteristics of the particular pilot. To ensure the integrity of the structure, inspections must be conducted on each aircraft to determine the existence of cracks at the critical locations. Rather than inspect each aircraft at specified usage or time intervals, it is more efficient to monitor the usage of each aircraft and make use of this in determining when inspections should be performed. In this way, inspections can be minimized and conducted at times when most needed (see Fig. 2-24).

Strain gages are used to monitor peak strain conditions, and counting accelerometers are used to keep track of the acceleration peaks. Flight parameters such as angle of attack and Mach number are recorded for each flight of the aircraft. All this information is used to determine the differences between the baseline spectrum and how the aircraft is being flown in service. Flight parameters provide a backup method to the direct process of measuring strain and acceleration peaks. The baseline spectrum produces a known crack growth behavior at each critical point and, hence, a baseline life for each of those points. Calculation of the crack growth under the actual flight spectrum produces a projected crack growth life which is then compared to the baseline life for each critical point. The ratio of baseline life to projected life is called the usage factor. The usage factor can then be used to determine at what times in the actual life of the aircraft inspections should be done. For example, if the requirements call for an inspection at half of the design lifetime, the usage factor for a particular aircraft can be used to determine when that should occur in terms of real hours on the aircraft. Depending on the severity of usage, it may occur earlier or later than half of the baseline life.

Historic information about the relationship of certain flight regimes to structural damage are used in two ways. First, during the design of an aircraft structure, care is taken to ensure that the structure has the ability to withstand the projected load spectrum. Frequently, the historic data points up certain portions of the flight regime that cause specific structural difficulties. The structure can be designed to withstand these conditions. Second, during the operational life of the aircraft, the fleet commander placards (places an upper limit on) the usage of the fleet to minimize risk, as schematically shown in Fig. 2-24. This is of benefit when unexpected failures occur and decisions are required about the continued operation of the fleet until the full extent of the problem is assessed. Continued operation at a reduced usage level frequently reduces the risk of failure sufficiently, such that time is made available for complete assessment of the problem. Control of usage is also beneficial in avoiding routine operation in parts of the flight regime that are known to be particularly severe for some failure modes.

These principles can be applied to other applications. For example, although every nuclear power plant is unique, there are similarities in the basic reactor systems design from one plant to the next. This results in similarities in the behavior of systems. Historic information can be used to provide an awareness of problem areas. This is again useful in the design of plants as well as in the operation, maintenance, and inspection of existing plants. The design can be sensitive to problem areas that have been documented historically. In addition, the historic information can be used to determine a preliminary operating regime for the plant. Monitoring of critical locations within the plant allows the operational envelope of the plant to be fine-tuned to reflect the specific characteristics of the plant. In this way, general information from across the industry can be used to reduce risk, and the general information can be modified to take into account the behavior of the particular plant from the on-line monitoring of the critical locations.

Welds, for example, share common factors in geometry and material that allow large numbers to be grouped together with their primary differences being the load environment. The load environment must include temperature, pressure, and flow effects. (Recent work has been done in the aircraft industry to study the effects of combined thermal and mechanical loadings on crack growth. This could be coupled with the results found from studies in the nuclear industry.)

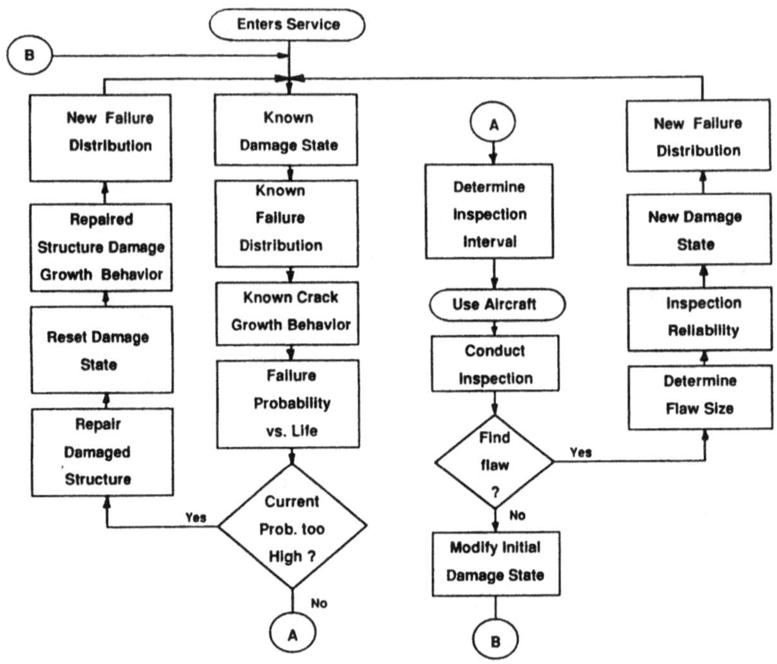

a. Individual Aircraft Application

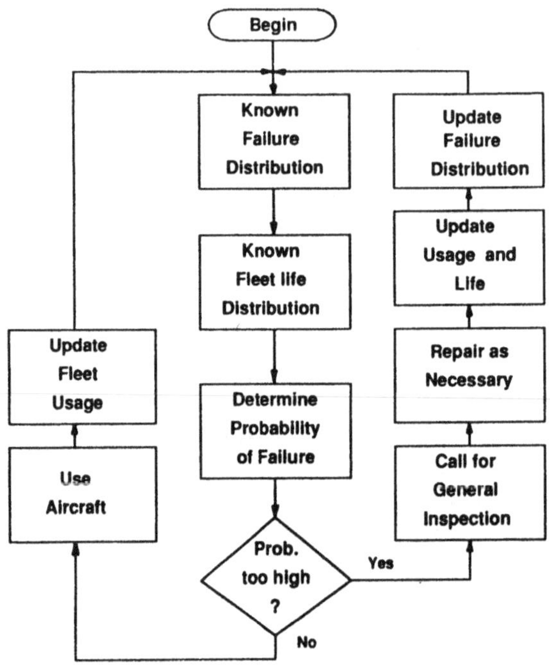

b. Application to Management of Fleet of Aircraft

FIG. 2-24 INSPECTION PROGRAM STRATEGY FOR INDIVIDUAL AIRCRAFT VS. FLEET MANAGEMENT EXAMPLE

Methods must be developed which allow a usage factor for each critical location in the plant to be determined from the operating parameters that are monitored and recorded as a matter of course. This is necessary as a backup to the direct methods of monitoring, as well as providing the operators with the ability to control usage should that become necessary from a risk standpoint. In addition, normal operation can be controlled to avoid conditions that are known to cause accelerated damage or greater risk of failure.

The plant operating profile in terms of the reactor conditions, power output, and others must be related to the crack growth conditions at the critical locations where structural monitoring is being conducted. This allows the development of an operating (or load) profile from which the usage of each critical location could be determined or controlled. As in the case of aircraft, the usage information is beneficial in determining when and where to conduct inspections. If detailed monitoring shows that a particular location is experiencing more severe operating conditions than previously anticipated, inspections are focused in this area to reduce the risk associated with unexpected failure.

A synergy exists between the historic information that provides a general backdrop for the design and operation of a plant, the detailed monitoring of pressure, flow, and temperature at specific critical points that provides the unique behavioral characteristics of a given plant, and the plant operating parameters that are controlled by the operators. These all come together to provide a basis for establishing inspection strategies, along with controlling operation, and thereby anticipating and averting problems rather than just reacting to incidents that occur unexpectedly.

Define Potential for Damage States and Inspection Damage. (See Fig. 2-25.) Section 2.3.3 of this document provides insights into determining the potential for damage states to exist in a given system, component, or element.

Relative to defining the potential for damage (or personnel danger) to occur from a defined inspection strategy, the following aspects should be considered, as a minimum:

- exposure of inspection personnel to a threatening environment, including hazardous material; radiation; excessive dust; extreme temperatures, pressures and wind; dangerous heights or climbing of scaffolds and unsteady platforms; rotating equipment or machinery; and falling objects
- degree of success in placing back together systems, components, or elements that have to be taken apart or taken off-line to do the inspection, including the potential for loose parts to occur
- movement of large equipment or structures that can damage adjacent equipment or structures

Define Reliability of Inspection Methods (See Fig. 2-26.). The effectiveness of alternative inspection methods can vary widely, and it is therefore important to consider inspection reliability in the development of an inspection program. This section presents examples of data that quantify the effectiveness of some typical inspection methods, with the parameter "Probability of Detection (POD)" used as the measure of inspection reliability.

An inspection method incorporates the aspects of procedures, tools, and level of personnel qualification and training, all of which impact the ability to detect the structural degradation of concern for the component. Inspection tools can range widely; for example, Fig. 2-27 shows early work by Packman (1968) that developed a reliability index for characterizing flaws in aircraft structures for various nondestructive examination (NDE) methods. Some inspections may require only purely visual examinations, whereas in other cases, the application may call for "high technology"

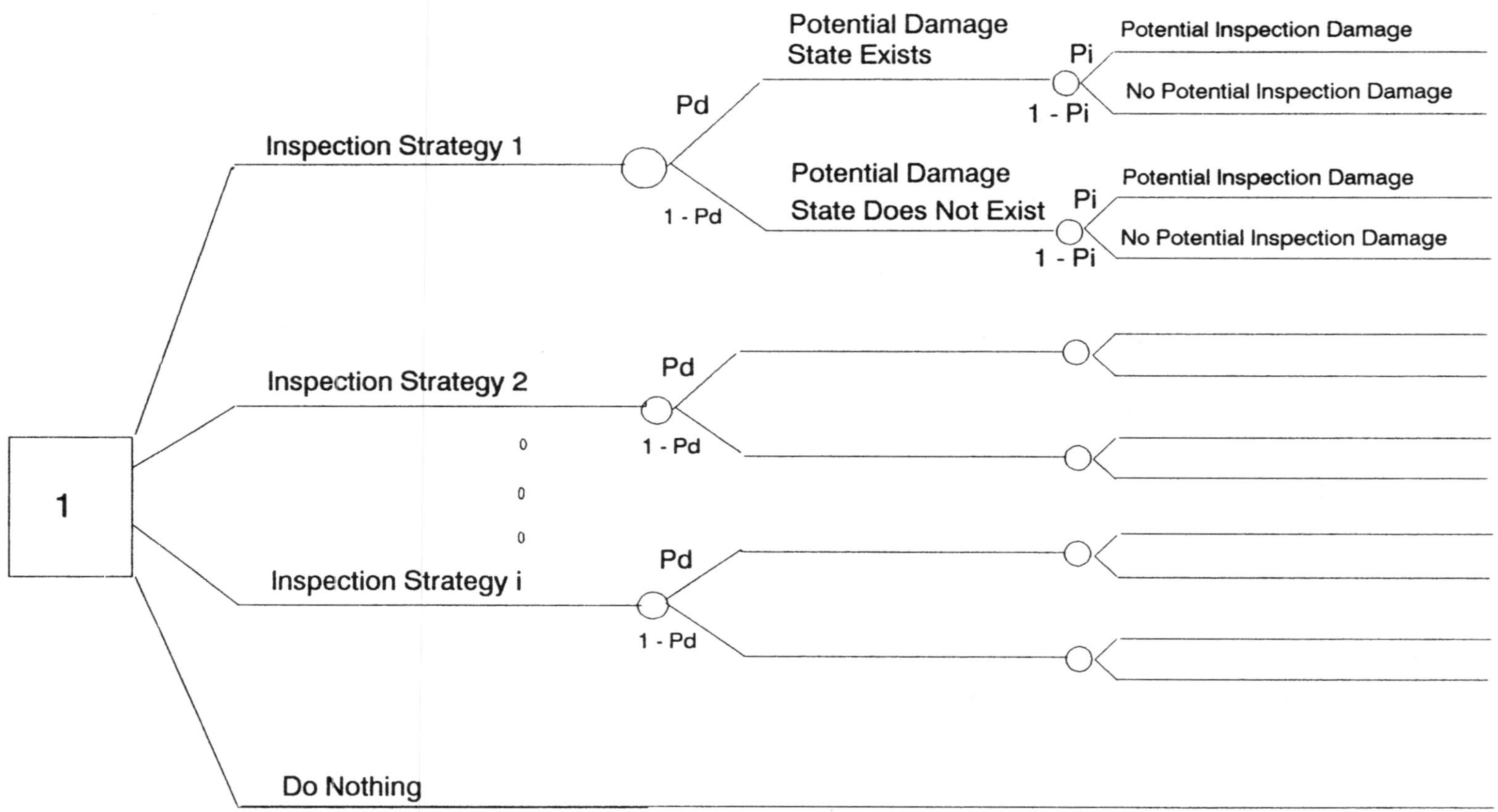

FIG. 2-25 SKELETON DECISION TREE FOR CHOOSING CANDIDATE INSPECTION STRATEGIES (DECISION NODE 1 AND CHANCE NODES FOR POTENTIAL DAMAGE STATES AND INSPECTION DAMAGE)

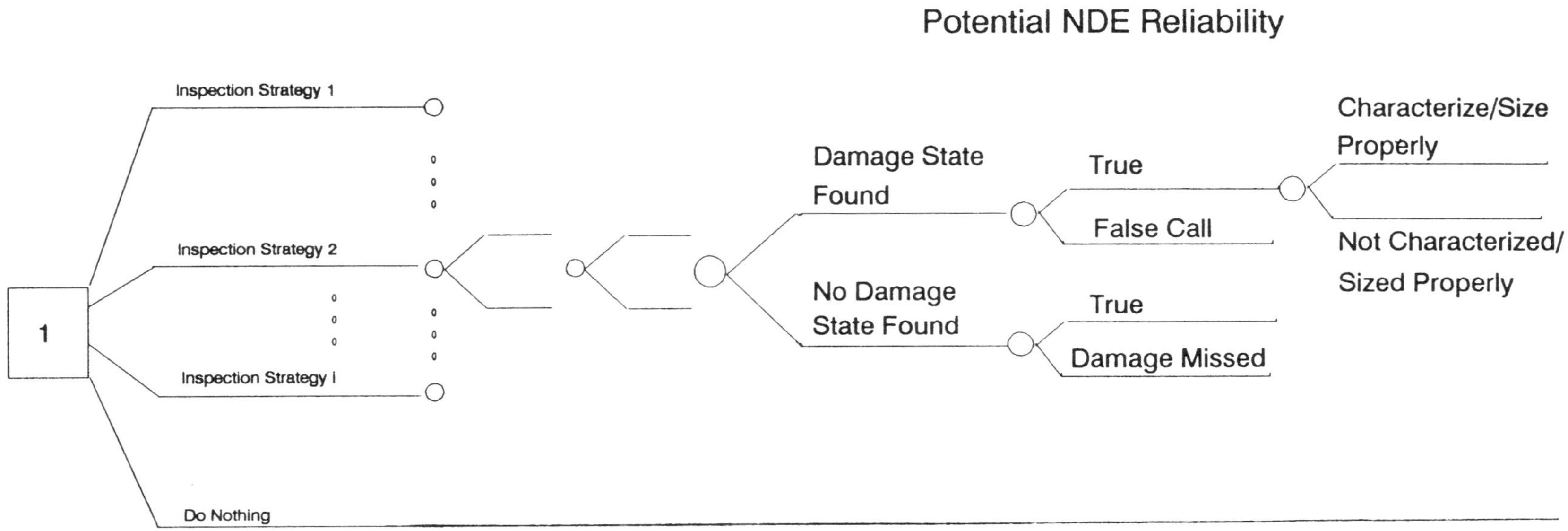

Note: Prune Tree and Define Candidate Inspection Strategies From Results

FIG. 2-26 SKELETON DECISION TREE FOR CHOOSING CANDIDATE INSPECTION STRATEGIES

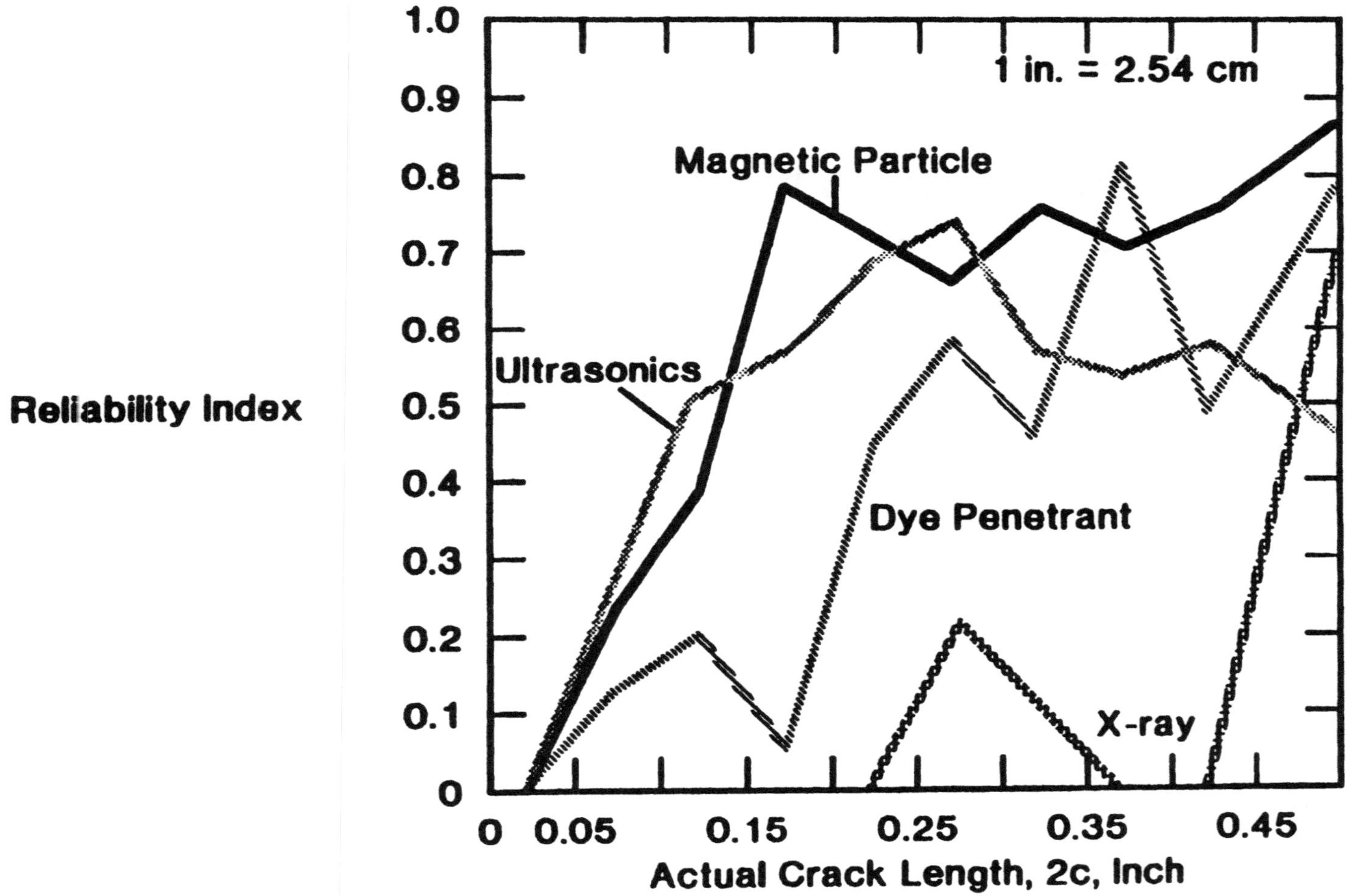

FIG. 2-27 COMPARISON OF FOUR NONDESTRUCTIVE EXAMINATION TECHNIQUES ON RELIABILITY OF DETECTING FLAW INDICATIONS IN STEEL CYLINDERS
[Packman (1968)]

ultrasonics to detect small cracks buried beneath the surface of the component or structure. In many cases, a combination of two or more methods may be the most effective strategy.

It is important to select an appropriate inspection method, since the most sophisticated technology may not always be needed for an effective inspection strategy. For example, a visual inspection can give a very reliable detection of general corrosion, provided that the inspections are performed on a timely basis and the inspection procedures permit visual access to all surfaces of the component that are subject to corrosive attack. A visual walkdown can also detect leaks that may be occurring.

The selection of inspection methods should be closely coordinated with requirements for inspection frequency. In many cases, frequent inspections of lower sensitivity can provide a greater impact on the integrity of the structure than can be obtained by a more sensitive inspection, which can only be rarely performed during the service life of the component. On the other hand, inspections with very low detection capability are of little value since they only create a false level of confidence in structural integrity and result in the diversion of limited resources from other more worthwhile inspections.

The sensitivity and reliability of NDE methods (in particular ultrasonic examinations) have been the subject of research in the Programme of Inspection for Steel Components (PISC)[3] (1986) and by Taylor, et al. (1983) and Doctor, et al. (1983) for the U.S. Nuclear Regulatory Commission. Extensive databases have been generated from round-robin type inspection trials, and this data has been reduced to statistically based curves for probability of crack detection. Such curves can be very useful in evaluating the potential benefits of alternative inspection plans.

Figure 2-28 shows recent results from Doctor, et al. (1990) for the detection of cracks in reactor pressure vessels that are obtained from inspection trials performed under the PISC-II program. The detection capability is best for the larger defects and approaches 100% when the very best procedures are utilized. With less sensitive procedures, the detection probabilities are greatly reduced, down to levels of less than 50% in some cases.

Figure 2-29 summarizes data from several sources [Harris, et al. (1981), F. A. Simonen and Woo (1984) and F. A. Simonen, et al. (1986)] for the detection of crack-like flaws in ferritic steel components such as reactor pressure vessels and thick-walled piping. While all curves of Fig. 2-29 show increased ultrasonic detection capability (approaching 99%) for larger cracks, all cases show a 50% probability of missing relatively small cracks of 0.1 in. in depth. The large spread in the detection curves indicates the variability and uncertainty in estimates of NDE reliability. The variability is in part due to significant differences in the inspection methods, but is also strongly dependent on the characteristics of the component being inspected (e.g., tight vs. open cracks, rough versus smooth surfaces at the interface between the transducer and component, material type, etc.).

POD must also be addressed in conjunction with the concept of false call probability (FCP). In Fig. 2-30 from work by Doctor, et al. (1988), the data applies to ultrasonic detection of stress corrosion cracks in stainless steel pipe weldments, which are relatively difficult to examine by ultrasonics. The figure depicts the performance of six inspection teams examining a common sample of test specimens. Curves result from different interpretations of data (i.e., threshold level of response signal needed to call a marginal indication a crack).

---

[3]The PISC research effort is conducted under the "umbrella" of the Nuclear Energy Agency of the Organization for Economic Cooperation and Development and the Commission of European Communities (CEC). Countries in addition to the CEC, such as the United States of America, Japan, and others, are full members and participants in PISC.

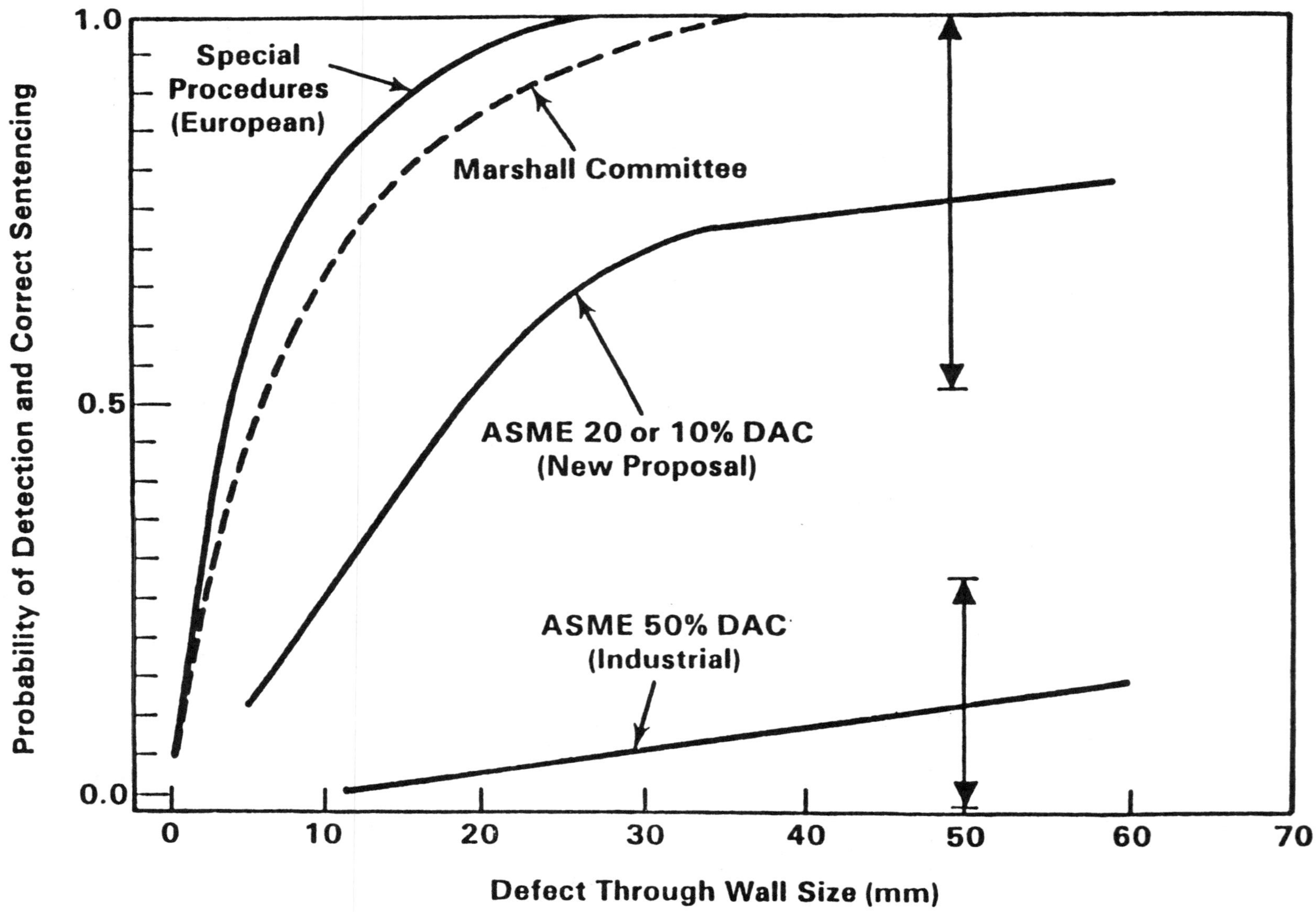

FIG. 2-28 PISC II OF ROUND ROBIN INSPECTION TRIALS FOR SELECTED PROCEDURES FOR PLATE NO. 3
[Doctor, et al. (1990)]

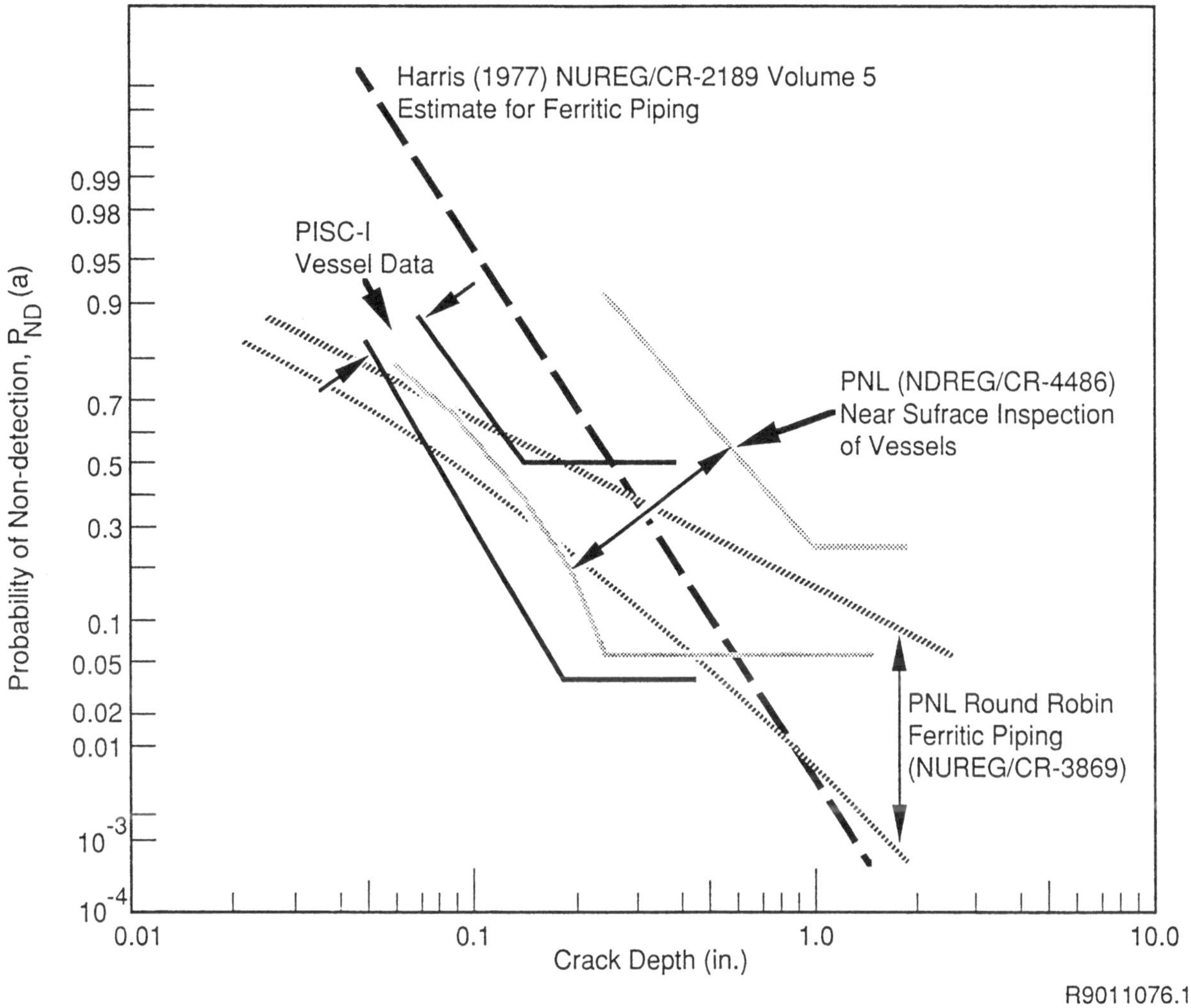

FIG. 2-29 PROBABILITY OF NONDETECTION OF A CRACK AS A FUNCTION OF ITS DEPTH FOR AN ULTRASONIC INSPECTION [ASME Code Section XI (to be published)]

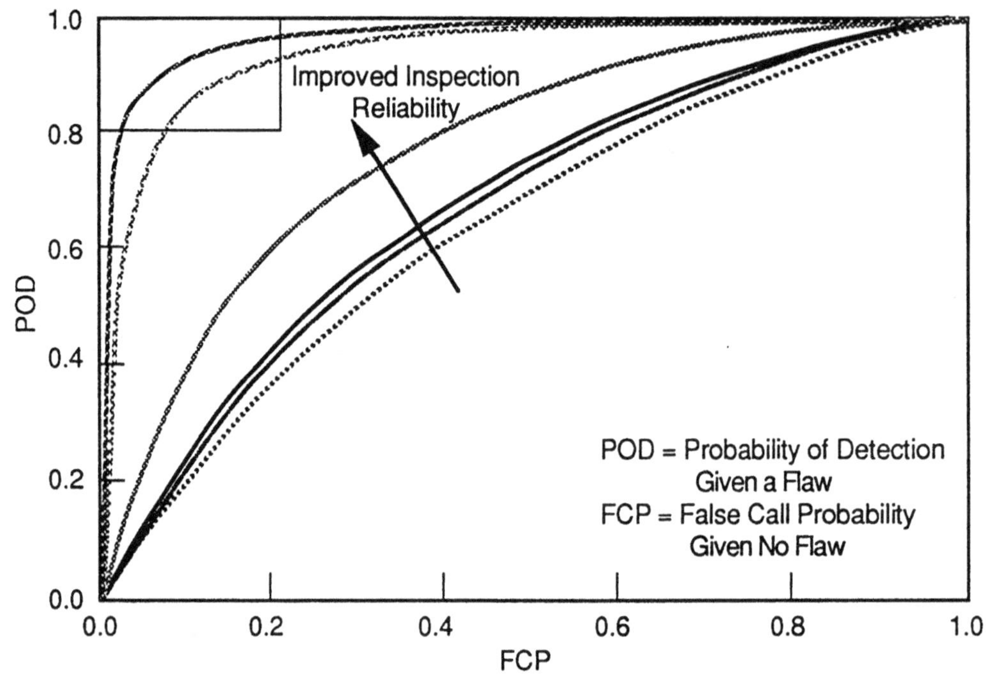

Note: Weld Cracks in Wrought Steel Piping

FIG. 2-30 VARIABILITY IN INSPECTION PERFORMANCE CURVES SHOWING INSPECTION PERFORMANCE OF DIFFERENT TEAMS [Adapted from Doctor, et al. (1988)]

There are clear differences in the performance achieved by the different teams of inspectors. A given team can also achieve various balances between improved detection and false calls by changing their judgmental criteria used to interpret the signals obtained from flaws during the ultrasonic examination. Human factors, which impact the discrimination of flaw signals, also play a significant role in inspection performance. Clearly, the poorer teams of Fig. 2-30 typically detected relatively few cracks while, at the same time, making many false calls by "detecting" many nonexisting cracks in the specimens. The best teams are capable of an "acceptable" level of performance as defined by the region of the upper left-hand corner of Fig. 2-30 within which the PODs are greater than 80% and the corresponding FCPs are less than 20%. The best teams can also approach an ideal 100% detection capability, but only at the expense of increasing levels of false calls, which can lead to the unnecessary repair or replacement of a component.

While the above discussion has focused on the detection of defects, an equally important consideration is that of defect sizing and, hence, sentencing. If one is still seeking to minimize the probability of significant defects entering service, i.e., maintaining a high integrity, then it is reasonable to assume that a plot not unlike Fig. 2-30 would again emerge. That is, one can only approach a low probability of allowing a significant defect to pass through, at the cost of an increasing number of unnecessary repairs for insignificant or, even at the extreme, no existing defects.

This does not present an issue for any theoretical analysis, but does have a significant impact on management. This impact comes from the implicit cost of these unnecessary repairs. If pressure is put on the inspection teams to drive down these costs, without giving them the necessary equipment and training, this goal will probably be achieved at the cost of plant integrity.

Research to improve the sizing and characterization of flaws is currently needed to improve the effectiveness of inspection. If undersizing of defects can be minimized, then there will be a reduction in unexpected service failures. On the other hand, there will be fewer cases of unnecessary repairs if oversizing of defects occurs less frequently.

In summary, the selection of candidate inspection methods is an important step in the development of any inspection program. These methods must be sufficiently reliable to meet the goals of the program. In addition, both the strengths and limitations of alternative methods need to be realistically assessed if cost-effective inspections are to be performed.

2.6.3 □ Choose an Inspection Strategy and Perform Inspection (See Fig. 2-31.)

Structural reliability and risk assessment (SRRA) is a process that can provide a best estimate of the probability of component failure, given various ranges of consequence, as a function of a presumed initial damage state, the reliability of the NDE method, the potential degradation mechanisms and associated loading conditions, and the life of the component or facility. The results from this effort can help in ranking components or elements for inspection and for evaluating inspection strategies.

For example, *probabilistic fracture mechanics* calculations can be a useful SRRA tool in relating an improved NDE reliability method to reduced failure probabilities for inspected components, compared to using traditional procedures. Figure 2-32 from E. P. Simonen, et al. (1986) shows that an increase in NDE reliability can reduce estimated failure probabilities for a reactor pressure vessel by a factor of about 10. The less reliable inspection (i.e., "old" ASME Code Inspection in Fig. 2-32) has only a modest impact on failure probabilities, and then only for the extreme case of an older vessel that has been significantly embrittled by a high level of neutron fluence.

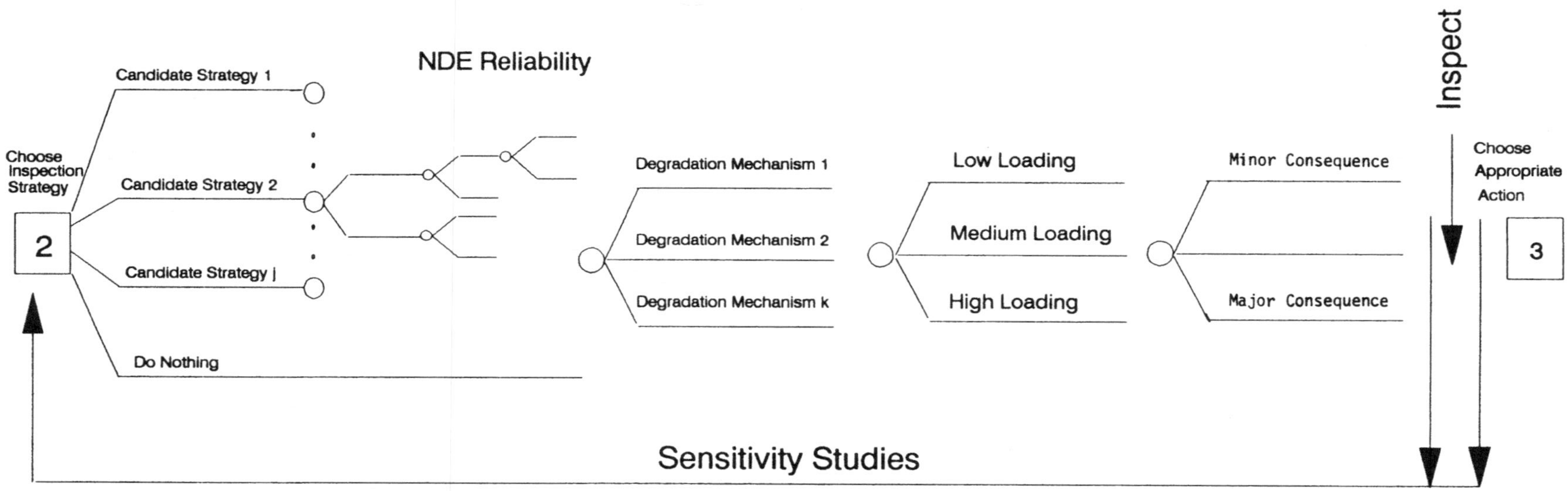

STEPS: 1) Perform sensitivity studies, as necessary, then select inspection strategy

2) Perform inspection

3) Perform sensitivity studies using inspection results, as necessary, prior to next decision

FIG. 2-31 SKELETON DECISION TREE FOR CHOOSING AN INSPECTION STRATEGY

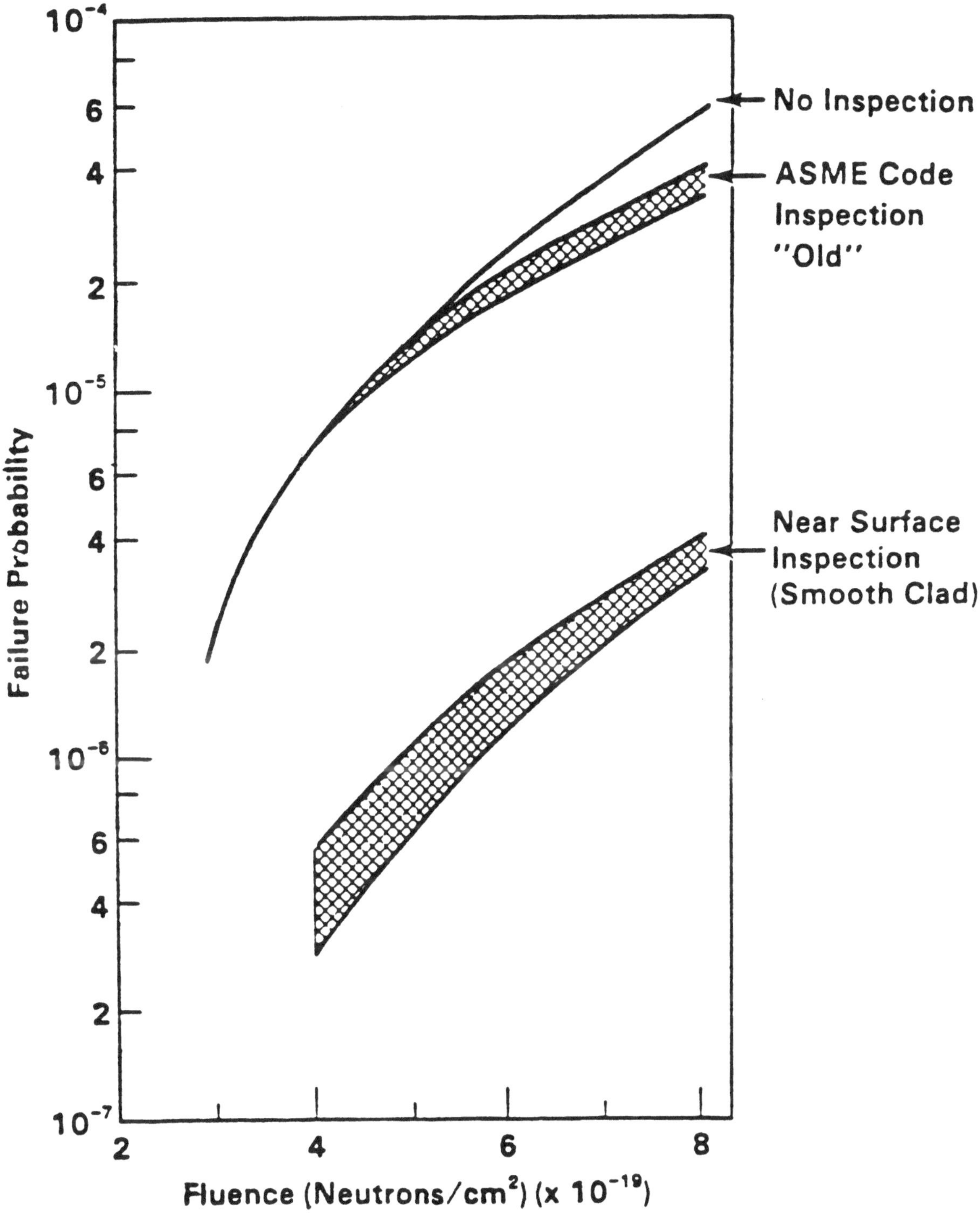

FIG. 2-32 IMPACT OF NONDESTRUCTIVE EXAMINATION ON PRESSURE VESSEL FAILURE PROBABILITY
[Simonen, E. P., et al. (1986)]

In Fig. 2-33 from F. A. Simonen and Woo (1984), the impact of inspection on component reliability is markedly different for the two situations, even though the method and frequency of inspection is exactly the same for both cases. Case A addresses a situation where the failures tend to occur early in the operating life of the component, and little reduction in risk is achieved by inspection since most failures occur prior to the time of the first periodic inspection (e.g., after 10 years of operation). Case B provides a contrasting example whereby significant benefits of inspection are achieved. In this case, most failures occur later in the life of the component, and thus are more readily detectable with the prescribed inspection frequency of once every 10 years. The benefits of inspection become increasingly apparent as the age of the component increases. Specifically, the inspection program provides an overall factor improvement of about 10 in reliability over the 40-year design life of the component. Figure 2-34, which is also from F. A. Simonen and Woo (1984), shows an extension of this philosophy to actual piping issues related to fatigue crack growth in carbon steel piping and intergranular stress corrosion cracking in austenitic stainless steel piping.

As mentioned earlier in the quantitative risk analysis section, SRRA results, if available, can also be used up front in the risk-based ranking process. However, the more important use of the SRRA process is in the incorporation of NDE results following an inspection. The SRRA process can be applied to assist in making repair, replace, or do-nothing decisions if damage is found during inspection and in updating the strategy for the next inspection. Following a summary of the SRRA process, considerations are presented for reusing the SRRA process following an inspection.

Structural Reliability and Risk Assessment Process. The benefits of inspection and resulting corrective actions in reducing the actual probability of failure are key inputs to a risk-based inspection process. Such benefits are estimated by the steps shown schematically in Fig. 2-35, which summarizes a general approach that considers structural degradation due to accumulation of damage. The rate of damage accumulation, the critical level of damage, and the probability of detecting damage by inspection are the required components of the procedures.

Examples of damage models include initiation and propagation of cracks (by mechanisms of fatigue, creep, or stress corrosion), general corrosion, fretting, and wear. Analytical procedures are probably most highly developed for the growth of cracks due to cyclic loading (fatigue), and the items in square brackets in Fig. 2-35 provide examples specific to a fracture mechanics analysis of fatigue crack growth.

The underlying procedures for analysis of damage accumulation can be based on deterministic models, with certain inputs being random variables, rather than deterministically defined values. An example is the initial crack size, which is seldom precisely defined, but cracks can be considered as being present with a probability that depends on their size. In principle, the failure probabilities are generated from the underlying deterministic procedure, considering some of the inputs to be random variables. A variety of procedures for obtaining actual results are available and have been used. Monte Carlo simulation is usually the most straightforward to implement, but may not be the most computationally efficient. Uncertainties in the values or randomness of some input variables often exist. Widely accepted means of treatment of such uncertainties are not available. The treatment of uncertainties, as discussed in Section 2.5.5 of this document, also applies here. Sensitivity studies are often useful in treating uncertainties, which are discussed later in Section 2.6.4.

The effect of inspection enters into the problem through a consideration of the probability of detecting damage as a function of damage state. An important consideration is the action taken once damage is detected, as this has an important effect on the post-inspection damage state. The difference between the pre-inspection damage state and the post-inspection damage state is a primary factor influencing the benefit of

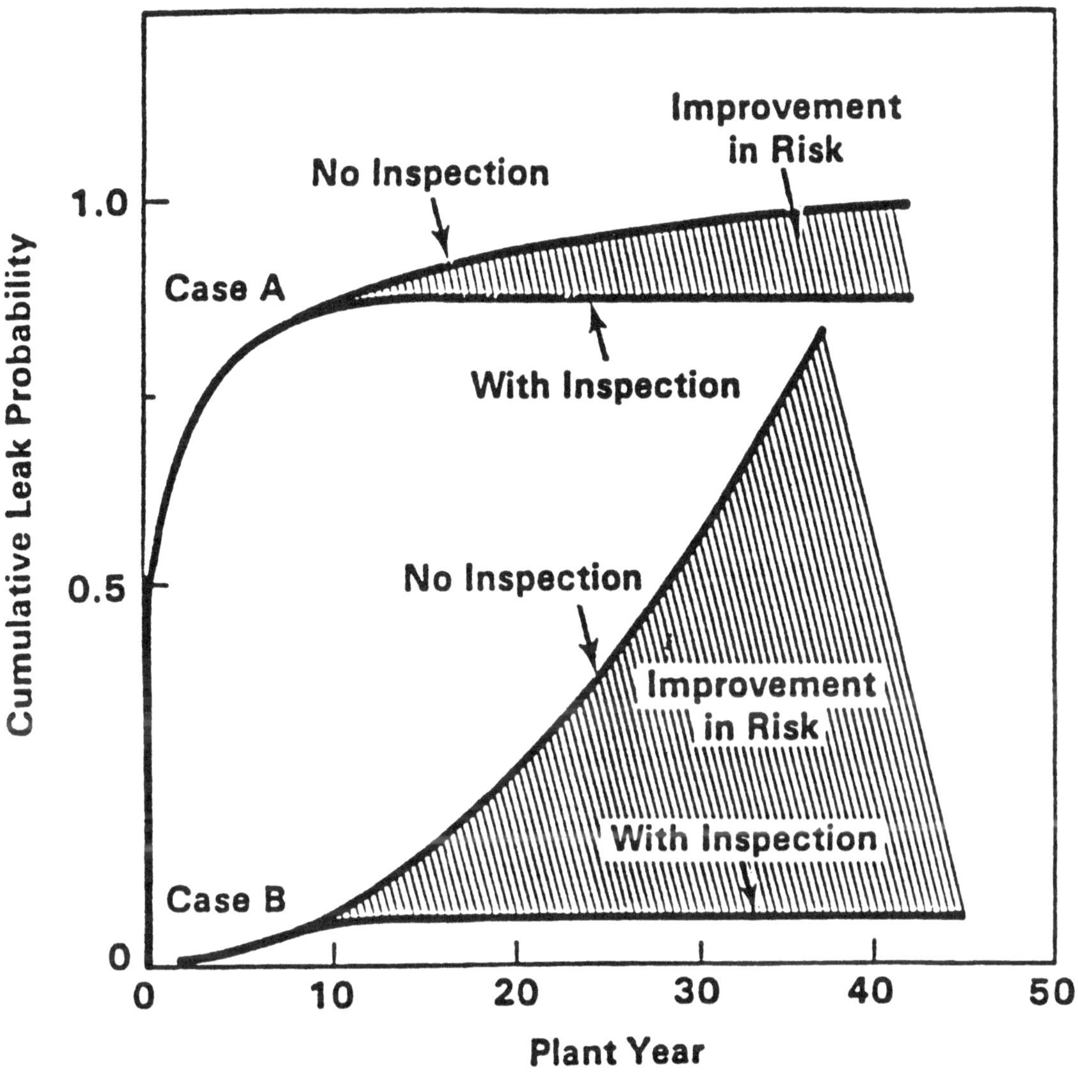

FIG. 2-33 IMPACT OF INSPECTION ON SYSTEMS WITH DECREASING (CASE A) VS. INCREASING (CASE B) FAILURE RATE [Simonen, F. A., and Woo (1984)]

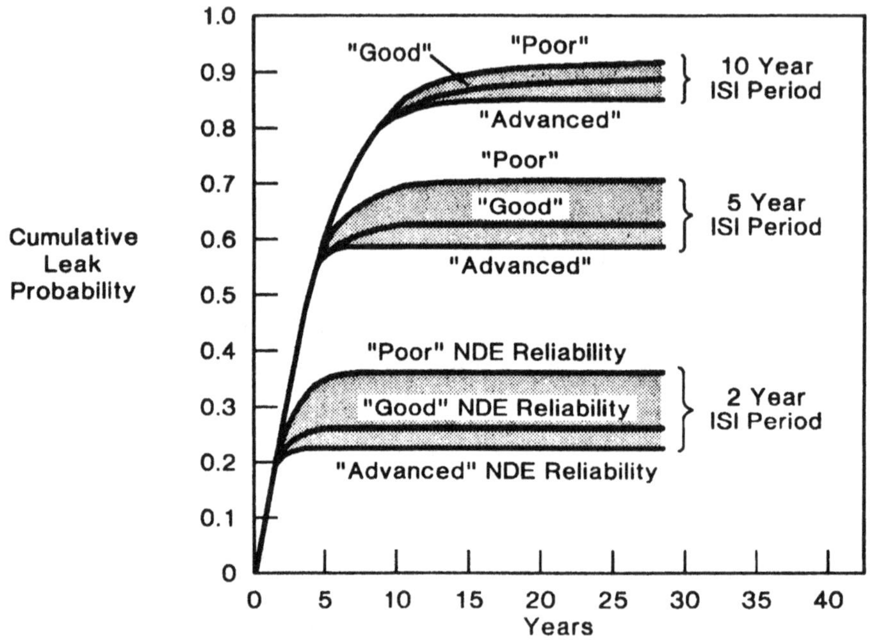

a. Thermal Fatigue of Pressurized Water Reactor Feedwater Line Nozzle

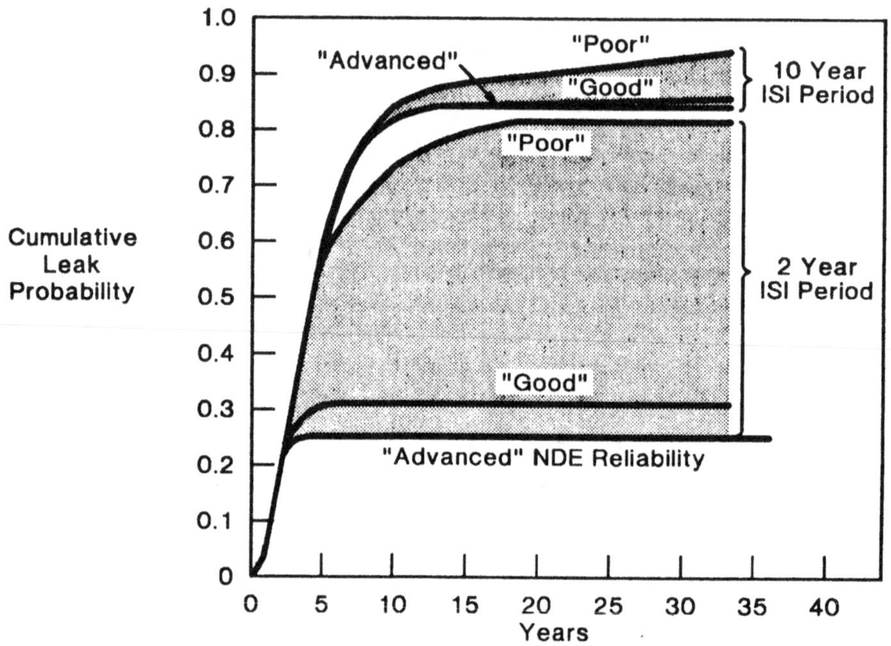

FIG. 2-34 EFFECT OF INSPECTIONS ON FAILURE PROBABILITY USING SRRA [Simonen, F. A., and Woo (1984)]

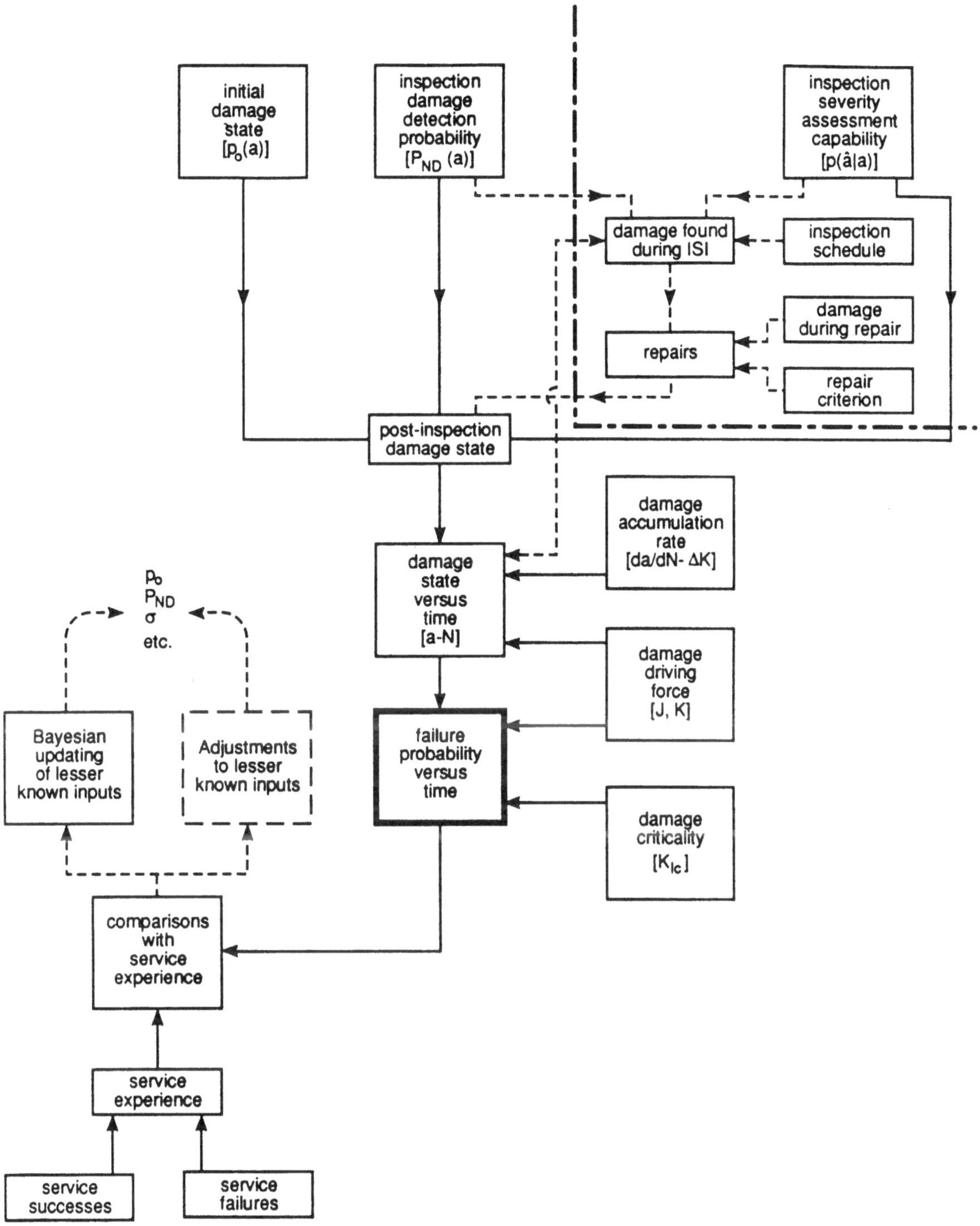

FIG. 2-35 STRUCTURAL RELIABILITY AND RISK-ASSESSMENT PROCESS

inspection (i.e., the change in failure probability in future operation due to the inspection) and associated corrective action.

Fairly general results for the effects of inspection on failure probabilities are generated from a "first principles" approach as described above. In many instances, there may be voluminous past experience on a specific component which may also provide information on the benefits of inspection. In such instances, results from past experience are all too often lacking for a specific problem of interest, and a combination of "first principles" and past experience provides a better approach to the assessment of inspection benefits.

Figure 2-35 shows schematically a comparison of "first principles" failure probability predictions with service experience. Such comparisons provide "calibration" of the "first principles" model, thereby increasing accuracy and confidence in predictions. Such calibrations consist simply of adjustments to lesser known inputs, or *Bayesian updating* of portions of the model.

The above discussion is an example of one approach to the assessment of the benefits of inspection. Other approaches are available. The above discussion considers the benefit of inspection to result from the actual detection of damage and its subsequent removal and replacement with undamaged material. This occurs with a probability that depends on the statistical distribution of damage present at the time of the inspection and the probability of damage detection as a function of damage state. An alternative approach that is discussed in the following sections is to consider the benefit of inspection to follow from the information that it provides concerning the state of the element. If it is inspected more often, then we know more about its state, which increases our confidence that we can evaluate its reliability. This increases our confidence that we can assess its capability to perform its intended function.

Once the effects of inspection/corrective action on the failure probability are available, the benefit of inspection can be assessed by comparing the cost of the inspection with the corresponding cost of failure. Models such as discussed above allow various inspection procedures and schedules to be studied in search of the optimum procedure that minimizes costs.

2.6.4 □ Choose Appropriate Action and Update State-of-Knowledge (See Fig. 2-36.)

Following an inspection, a decision must be made regarding a redefinition of the inspection strategy or taking corrective action if damage is found. However, before any decisions are made, it is recommended that the SRRA process be exercised whether or not findings occur (refer to the logic diagram in Fig. 2-31) in order to better determine the confidence in the level of safety. As shown in Fig. 2-36, the SRRA process should clearly be exercised in order to evaluate appropriate corrective actions if damage is found. This sensitivity study is discussed next, including the impact of having to take corrective action, and followed by recommendations for updating the state-of-knowledge, regardless of what path is followed.

Sensitivity Study Following Inspection. Inspections in themselves do not affect the probability of failure for a component or element to demonstrate an acceptable risk. Rather, inspection data provides a means of building up confidence in the level of safety of the component being inspected. The value of the inspection is measured directly with the SRRA process, as shown by Chapman (1989). There remains, however, the basic question of how much confidence can be placed in the SRRA predicted failure probabilities, and, hence, in the safety justification. The objective of the inspection policy must, therefore, be to build confidence in the original probability analysis. This implies that the effect of the inspection on the probability of failure is not

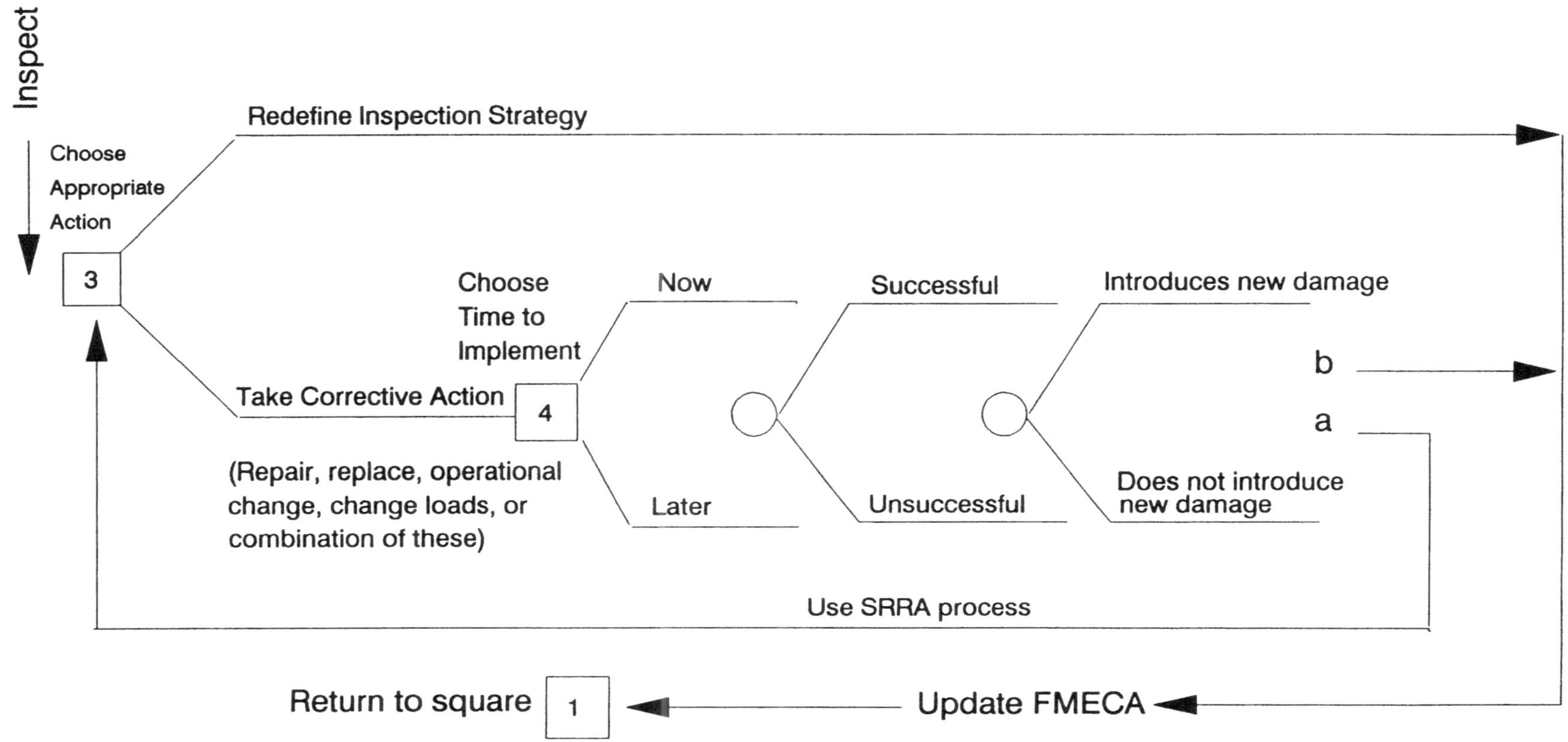

FIG. 2-36 SKELETON DECISION TREE FOR CHOOSING APPROPRIATE ACTION

the key concern. True, any inspection must affect the probability, but if a sample inspection is to be put forward, it is doubtful that this will significantly affect the overall plant safety. For example, if 80% of the risk is contained within the sample, and here it is assumed that the ranking has correctly identified this sample, then reducing the probability of failure by two decades for this sample only gives a factor of 5 improvement to the overall probability of failure. Despite the above, it would be imprudent not to inspect the high-risk areas. Furthermore, these areas are the ones most likely to give a positive outcome. But, the basic justification for a sample size inspection must lie in its role as a confidence-building exercise and not a failure rate reduction procedure.

At this point, SRRA is reused to investigate the effect of changing the basic assumptions originally used to build up the failure probability. A statistical method, as given in Appendix C from Chapman (1983), can be used to evaluate the benefits of inspection for future strategic decisions. The method is based on Bayesian logic using the information obtained from inspection on a particular family of components to update a vague prior belief about that family. The vague prior belief consists of a collection of what are called "probability failure sets" (PFSs). Each PFS consists of separate sets of defect density, defect size distribution, inservice crack growth, and critical crack size. From this basic data, a specific probability of failure is determined that is uniquely linked to each PFS. The basic data in each PFS is now used to predict what happens to the defects as a function of time, and, hence, what results would be expected from a given inspection program of work. It is the comparison of these predicted outcomes with the actual outcome that is used to identify which of the PFSs is most likely to be true. For example, if no defects are found during inspection, the method builds up confidence that the true probability of failure for that family is very low. If defects are found, the method indicates which of the sets is most likely to be correct, and, hence, what the true distribution of failure probability is for that family of components. The changing probability associated with each PFS is now used as an indicator as to the reasonableness, or otherwise, of original analysis, which is discussed further by Chapman (1985).

The PFSs must cover a wide range of probabilities, from the most optimistic to the most pessimistic; a range which must, by definition, encompass the best estimate (if it exists) and the target or acceptable probability of failure. A distinct advantage inherent in this is that it is not necessary to be able to identify what the actual, or true, situation is; only that it lies within the range of the chosen PFSs. It is the inspection data that is used to try to identify the true situation, as shown in Fig. 2-37.

The model can be used retrospectively on inspections carried out to date. It can then be used as an ongoing method of assessing the most likely true situation regarding the probability of failure of any given family of components, and thus direct future inspection programs in the most effective way.

Since the model assumes that a given family of components is amenable to a statistical interpretation, it is necessary only to consider a sample of that family to provide information about the whole. This in turn implies that only a limited or reduced inspection program is required to provide the necessary information. It should, however, be clearly understood that if the outcome of a limited inspection program indicates that the probability of failure is too high, increasing the number of inspections or the sample size per inspection does not necessarily change the situation. The inspections provide more data which increases the confidence in the failure distribution, but this may only serve to confirm that the situation is unacceptable. If this becomes the situation, the basic premise of the analysis must be changed to ask how does inspection affect the probability of failure? It would be possible to develop the model toward this objective.

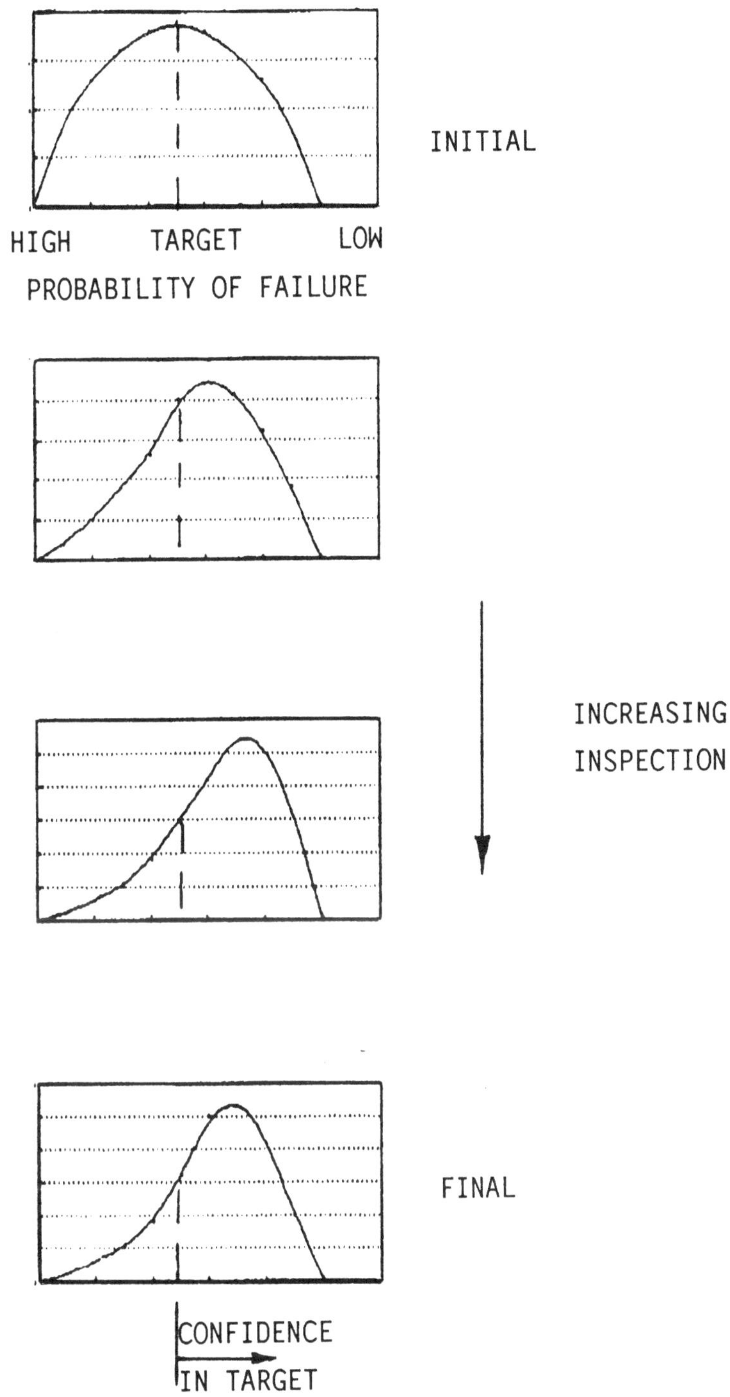

*Situation shown is for inspection results that show few defects of relatively small size

FIG. 2-37 DISTRIBUTION OF PROBABILITY FAILURE SETS*

The model does not implicitly include any form of consequence ranking. This can only be applied to the model by designating different acceptable levels for the failure probability for different family groups. But here, it must be realized that the model attempts to determine the true situation and cannot therefore be artificially made to meet a given target, without corrupting the inspection results input to the model. Thus, it must be concluded that the question of consequence is external to the model, but can be included in the analysis. If the model is to be used to plan and optimize a future inspection program, assumptions about the outcome of the inspection must be made. In this context, the consequence of failure must be used in order to determine the probability of failure that is acceptable to any given family of components. The model can then be used to optimize the inspection program against a given confidence level for the prescribed failure probabilities. Note that this does not mean that the confidence level is reached, since this depends on the realization of the inspection program. It does, however, provide a meaningful way of planning.

The method of analysis proposed in Appendix C is still under development. It is, however, at a point where it can be used to provide initial estimates of the benefits of inspection and to plan a future inspection program. The future development will not change the basic principles laid out in this report. The authors believe that the proposed method provides a meaningful way of interpreting the findings of an inspection program and making decisions based on the results.

If the above process shows that corrective action is required, the same process is applied. After potential corrective actions are identified and basic questions regarding the time of implementation, the success in mitigating the damage, and whether or not new damage may be introduced, the basic assumptions can be altered in the SRRA process to evaluate the effect of the corrective action. For example, crack growth rates may be drastically altered if new components with different material are being introduced for replacement. Experience in repairing aged structures shows that more uncertainty is usually introduced with corrective actions since little data generally exists relative to the structural analysis input assumptions. This condition emphasizes the need for the use of the SRRA process to address these uncertainties before making a selection. One corrective action or combinations of corrective actions may be chosen.

Update State-of-Knowledge. The above results are now used to update the FMECA information to rerank the components or elements based on risk. The risk evaluation plots given in Figs. 2-6a through 2-6e or in Figs. 2-7a through 2-7e are now revised. Figure 2-38 exemplifies an update to Figs. 2-6c and 2-7c, respectively. From these results, the inspection strategies are redefined (see Fig. 2-22), making a "living process."

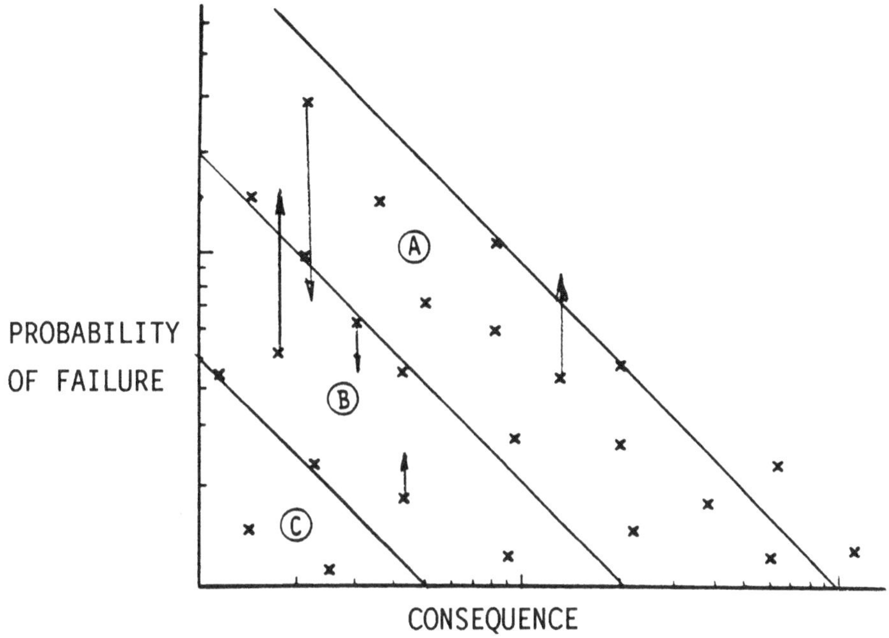

a. Ranking Using Single Point Estimates Based on Lines of Constant Risk

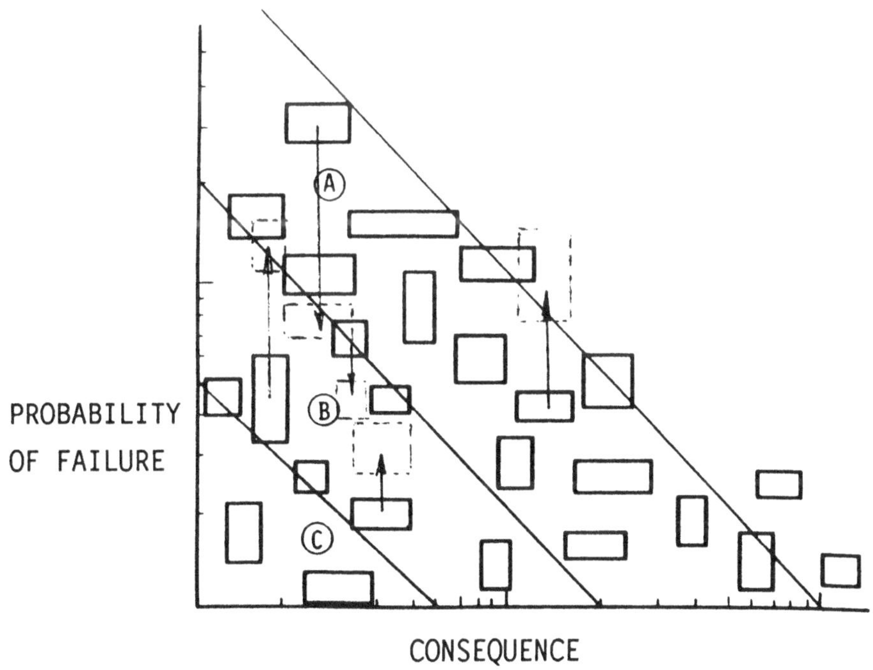

b. Ranking, Including Uncertainty, Based on Lines of Constant Risk

FIG. 2-38 RISK RANKING (POST INSPECTION)

DEFINITIONS[1] 3

accident sequence – a combination of events leading from an initiating event that challenges safety systems to an undesired consequence. The sequence is ordered, starting with the initiating event, through sequential failures leading to the consequence. Event tree analysis is often used to identify and quantify the probability of accident sequences. Dominant accident sequences are those associated with the highest risk (probability times consequence).

Bayesian updating/Bayesian logic – a process of updating or modifying an assumed probability distribution for some parameter (e.g., crack size distribution) using information from events which depend on the parameter (e.g., inspection results and/or fractures). The updating calculations use equations derived by the Reverend Thomas Bayes in 1763.

consequence – the impact (results) of a structural failure or release (i.e., gas cloud, fire, explosion, evacuations, injuries, deaths, environmental impacts) and the ultimate results of the structural failure or release in terms of impacts on public and employee health and safety, impacts on the environment, costs or damage to the facility, or loss of capability to produce a product.

cut set – a particular combination of failures of specific components due to specific failure modes, which yields an accident sequence leading to an undesired consequence. Generally, there are many cut sets associated with each accident sequence, since a variety of component failure combinations may result in the same loss of system functions. A minimum cut set is one having the minimum number of failures necessary to lead to the undesired consequence (i.e., having no redundant failures). Generally, only minimum cut sets are analyzed, and the word minimum is often not stated.

decision analysis – a prescriptive methodology for improving the chances of making rational choices in the face of uncertainty and risk. The methodology consists of four sequential stages: (1) problem-structuring, (2) deterministic systems modeling, (3) probabilistic analysis, and (4) evaluation. Influence diagrams and decision logic trees are generally used to construct and execute the process.

deterministic analysis – an analysis approach that uses bounding calculations and safety factors to deal with variability and uncertainty. The conventional engineering approach to analysis is deterministic, in which the result of a calculation is compared with an established acceptance criterion.

dispersion analysis – the evaluation, by means of a computer model, of the ambient concentrations of a hazardous material after its release. It takes into account physical and chemical states and properties of the hazardous material and the geographical, topographical, geological, and meteorological characteristics of the environment. It also considers the physical characteristics of the hazardous material, which influence

[1]Adapted from Lercari (1989) for the State of California Office of Emergency Services and from the U.S. Nuclear Regulatory Commission NUREG-1050 Report (1984).

the migration, movement, dispersion, or degradation of the hazardous material in the environment.

external events or external forces – events resulting from forces of nature, acts of God or sabotage,,or such events as neighboring fires or explosions, neighboring hazardous material releases, electrical power failures, earthquakes, and intrusions of external transportation vehicles, such as aircraft, ships, trains, trucks, or automobiles. External events are usually beyond the direct or indirect control of persons employed at or by the facility.

event tree analysis – an analytical tool which organizes and characterizes potential accidents in a methodical manner. The analysis process begins with the identification of potential initiating events that challenge facility safety systems.

Functional event trees are developed for each type of initiating event. These trees display safety functions which need to be provided after occurrence of an initiator to maintain the facility in a safe condition. When the functional relationships have been established, the functional event trees are developed into systematic event trees. That is, the events are expanded to depict the facility response in terms of the specific systems that perform the safety functions. The systemic event trees define specific system responses and requirements that are then used as top events for the system fault tree analyses.

facility – all buildings, equipment, structures, and other stationary items that are located on a single site or on contiguous or adjacent sites.

failure modes, effects, and criticality analysis (FMECA) – a specifically designed analysis to identify the conceivable failure modes of each component and the impact of the failure on operations, the system, and surrounding components. Failure modes and causes are determined for the purpose of inspection ranking from design information, operating experience databases, prior inspection results, prior risk-analysis results, and expert opinion from those with experience in operations, maintenance, and inspection, etc.

The FMECA begins with a qualitative analysis and is translated into a quantitative analysis when probabilities and consequences are estimated. Results of an FMECA yield data on each system component, failure modes, failure causes, impact on other components, and consequences. The FMECA allows systems and components to be ranked based on the probabilities and consequences, i.e., risk.

An FMECA does not serve well to identify multiple failures that lead to incidents.

fault tree analysis – a fault tree is a model of a system that shows the logical relationship between the elements of the system, and how failures of these elements can cause failure of the system. Fault tree analysis is the technique of deducing these interrelationships starting from specification of the system failed state, determining subsystem failure combinations that will cause system failure, and progressing stepwise into a more and more detailed specification of what can cause subsystem failure, until the failures identified are of basic components which need no further subdivisions. A fault tree is a qualitative model of the system until probabilities are assigned to the component failures involved; then the model becomes quantitative and system failure probability can be calculated.

frequency – probability over a specified time period.

hazard – a physical condition or a release of a hazardous material that could result from component failure and result in human injury or death, loss or damage, or environmental degradation. Hazard is the source of harm. Components that are used to transport, store, or process a hazardous material can be a source of hazard. Human error and external events may also create a hazard.

human error – any human action or omission which deviates from established or generally recognized acceptable procedures, the results of which could result in an event.

inservice inspection (ISI) – an inspection performed after preservice inspections and test runs are satisfactorily completed, and a facility, system, or component has been certified or accepted for normal service operation. The objective is to detect degradation induced by extended periods of operation.

interval estimate – a range of numbers with lower and upper limits that is assigned to some or all the parameters considered in the estimate to account for the uncertainty in these parameters.

likelihood – probability, frequency, chance, etc. For the purpose of the risk-based inspection process, likelihood is generally used with qualitative measures.

probabilistic fracture mechanics (PFM) – a specialized field of probabilistic structural mechanics related to calculating the probability of component failure from fracture.

probabilistic risk analysis (PRA) – an analysis that:

(1) identifies and delineates the combinations of events that, if they occur, will lead to a severe accident (e.g., major explosion) or any other undesired event;

(2) estimates the frequency of occurrence for each combination; and

(3) estimates the consequences.

The PRA integrates into a uniform methodology the relevant information about facility design, operating practices, operating history, component reliability, human actions, the physical progression of accidents, and potential environmental and health effects, usually in as realistic a manner as possible.

A PRA uses logic models depicting combinations of events that could result in severe accidents and physical models depicting the progression of accidents and the transport of a hazardous material to the environment. The models are evaluated probabilistically to provide both qualitative and quantitative insights about the level of risk and to identify the design, site, or operational characteristics that are the most important to risk.

PRA models generally consist of event trees and fault trees. Event trees delineate initiating events and combinations of system successes and failures, while fault trees depict ways in which the system failures represented in the event trees can occur. These models are analyzed to estimate the frequency of each accident sequence.

probabilistic structural mechanics (PSM) – an analysis methodology which combines deterministic analysis models with probabilistic representations of unknown or uncertain parameters. For example, random variations in loading conditions, flaw sizes, the mechanical properties and chemistry of structural materials, and the effect of degradation mechanisms may be represented by distribution functions which are combined to provide information on the reliability of the structure. Various methods for combining the random variables to estimate reliability are available, including Monte Carlo simulation, in which the variables are sampled randomly from their prospective distributions to obtain inputs for multiple simulations (deterministic calculations). The probability of failure is then determined from the fraction of simulations whose results indicate structural failure (e.g., unconstrained crack propagation).

probability – a mathematical basis for prediction that, for an exhaustive set of outcomes, is the ratio of the outcomes that would produce a given event to the total of possible outcomes.

release consequence analysis – evaluation of the potential health and environmental impacts from a release of a hazardous material. This is determined by dispersion analysis and comparison of projected concentrations of a released hazardous material with potential impacts. Historical events can provide some information on potential consequences. The hazardous material release rate, characteristics, dispersion rates, prevailing meteorological conditions, and unusual complicating factors (e.g., fire) are critical to this analysis.

risk – the measure of the potential for harm or loss (i.e., hazard) that reflects the likelihood (e.g., frequency) and severity of an adverse effect to health, property, or environment.

risk importance – a measure of the undesirability of an undesired consequence. Many risk importances have been defined, but they generally belong to two categories. The first category addresses the absolute or fractional risk to which failures or a given component (or system) contribute. The second category addresses the conditional risk associated with assumed failure of the component. Consider two components, each of which contributes 10% to the total risk (failure probability times consequences) of an operation. According to the first category, they have the same risk importance. However, the failure probability of one component is low and the consequence high, while the opposite is true for the other component. According to the second category of risk importance measure, the components have different risk importances. The component with the high associated consequences, given that it is assumed to fail, has a higher risk importance.

significant release – any release or potential release of hazardous material that poses a significant hazard to public health and safety or the environment. This concept includes releases at, or from, fixed facilities; covers employees; and includes transportation-related releases or potential releases.

single point estimate – representation of a variable that may have a distribution of values with associated probabilities. The single point estimate is usually a "best" estimate and is often a median value.

structural reliability and risk assessment (SRRA) – a method for evaluating component reliability, which takes into account information (e.g., failure data, analytical models such as for probabilistic structural mechanics analysis, expert opinion) or assumptions on the initial state of the component, degradation mechanisms, loading conditions, component lifetime, and the results and reliability of inspections. SRRA results can be used for risk-based prioritization of components for inspection, both prior to and subsequent to inspections. They can also be used in decision analysis in determining preferred actions (e.g., repair/replace vs. continue observation) if damage is found during inspections.

LIST OF SOFTWARE FOR RISK-BASED INSPECTION PROGRAMS

4

The computer software listed in Table 4-1 is a partial list of tools, some of which may not be commercially available, that the research task force suggests may be useful to those establishing inspection programs using risk-based information. Other tools are obviously available and should be applied accordingly to meet the needs of the user. It takes several tools in combination to carry out the entire risk-based inspection process. Additional software needs to be developed to fully implement and facilitate the processes that are recommended in this document.

TABLE 4-1
SUGGESTED SOFTWARE FOR RISK-BASED INSPECTION PROGRAMS

Purpose	Software Name	Source	Description
Database development	dBASE III PLUS*	Ashton-Tate, Torrance, CA	General software for establishing and sorting databases; useful for FMECA database generation
	Lotus 1-2-3*	Lotus Development Corporation, Boston, MA	General software that integrates spreadsheet, database and graphics; useful for risk-based ranking calculations
Probabilistic risk assessment	CAT	–	Constructs fault trees for simple systems using decision tables
	IRRAS, Version 2.0*	Idaho National Engineering Laboratory, Idaho Falls, ID	Uses fault/event trees as input; quantifications include: system and component failure rates, system and component prioritizations, and uncertainty analyses
	SARA, Version 2.0*	Same as above	Same as above
Statistical evaluations	MATH TOOL – Statistics I*	Gulf Publishing, Houston, TX	Performs statistical treatment of data to determine appropriate SRRA models, e.g., flaw and material property distributions
	STATGRAPHICS*	Statistical Graphics Corporation, Rockville, MD	Same as above
	UNIFIT*	Simulation Modeling and Analysis Co., Tucson, AZ	Same as above
Structural reliability and risk assessment – general application	PROBAN*	DNV Industrial Services, Inc., Houston, TX	Performs probabilistic analysis of general structures and components; can assist in definition of inpsection strategy
	SRRA*	Westinghouse Electric Corp., Pittsburgh, PA	Same as above
SRRA – Vessels	COVASTOL	CEA, Euratom, and Frametome, Paris, France	Performs probabilistic fracture mechanics analysis of reactor vessels for radiation embrittlement using histograms for simulation
	OCA-P	Oak Ride National Lab, Oak Ridge, TN	Same as above except uses Monte Carlo for simulation

* = PC-based software

TABLE 4-1 (CONT'D)
SUGGESTED SOFTWARE FOR RISK-BASED INSPECTION PROGRAMS

Purpose	Software Name	Source	Description
SRRA – Vessels	PFM	Westinghouse Electric Corp., Pittsburgh, PA	Same as above except uses Monte Carlo with importance sampling for simulation
	VISA-II*	Battelle Pacific Northwest Labs, Richland, WA	Similar to Westinghouse PFM Code
SRRA – Piping	PRAISE-B, PRAISE-CC*	Failure Analysis Assoc. Inc. (FaAA) for Lawrence Livermore National Lab, Livermore, CA	Performs probabilistic fracture mechanics (PFM) for analysis of piping systems for fatigue-crack growth (PRAISE-B) and intergranular stress corrosion cracking (PRAISE-CC) and includes the effects of inspection
	PRAISE-FBR	FaAA, Menlo Park, CA	Same as above except evaluates creep/fatigue degradation
	PEPC	FaAA, Menlo Park, CA	Performs PFM analysis of high-temperature piping under creep fatigue conditions
SRRA – Turbine	SAFER/PERL	Same as above for Electric Power Research Institute, Palo Alto, CA	Performs PFM analysis of steam turbine rotors for fatigue, creep–fatigue, and stress corrosion cracking
SRRA – Other	BOPPER	FaAA and Electric Power Research Institute, Palo Alto, CA	Performs probabilistic crack initiation and fracture mechanics crack growth for high-temperature piping and boiler components
	PERL	Same as above	Performs PFM calculations of lifetime of low-pressure steam turbine blades
	PACIFIC	FaAA, Menlo Park, CA	Performs PFM calculation of lifetime for wide range of crack geometries and stress systems; considers initial crack size, fatigue crack growth properties, and fracture toughness to be random variables
Decision analysis	ARBORIST*	Texas Instruments, Inc., Dallas, TX	Solves decision tree by averaging out and folding back the input data

* = PC-based software

TABLE 4-1 (CONT'D)
SUGGESTED SOFTWARE FOR RISK-BASED INSPECTION PROGRAMS

Purpose	Software Name	Source	Description
Decision analysis	DAVID*	Duke University, Durham, NC	An influence diagram processing system for the Macintosh
	INDIA*	Decision Focus, Inc., Palo Alto, CA	An influence diagram processing system for the Macintosh
	@RISK*	Palisade Corporation, Newfield, NY	Add in for Lotus 1-2-3 for performing Monte Carlo simulations
	Supertree*	Strategic Decisions Group, Menlo Park, CA	Performs decision analysis using decision trees for complex issues

* = PC-based software

SUMMARY AND RECOMMENDATIONS 5

A multi-disciplinary research effort has been performed to describe and recommend appropriate processes and methods using risk-based information for developing inspection guidelines for facilities or structural systems. In certain areas, these methods should be used to define risk-based inspection programs for recommendation to ASME and other codes and standards bodies. These methods should be applicable to all areas where structural failures have the potential to result in loss or damage.

A review of current inspection requirements and related research developments has been performed in order that an interdisciplinary approach for the risk-based inspection process could be developed by integrating technologies from a broad base of applications. Applications that have been explored range from nuclear and fossil fuel-fired power plant systems and components to aircraft, civil engineering, and marine ship hull structures. Applications of interest to the nuclear and conventional insurance industry have also been reviewed.

From this information, a four-part, risk-based inspection process is recommended to rank or classify systems, components, or elements for inspection and to develop the strategy (i.e., the frequency, method, and sample sizes) to perform the inspection. This process includes:

(1) a definition of the system;

(2) a qualitative risk assessment;

(3) a quantitative risk analysis that includes an enhanced failure modes, effects, and criticality analysis;

(4) the development of the inspection program for components and structural elements using decision risk-analysis methods.

This multi-disciplinary, top-down approach starts at the system level before focusing the inspection program at the component or structural element level. The strategy recommends that the results of inspections are continuously used to update the state-of-knowledge throughout the four parts of the approach so that a living process is achieved.

A key step in defining the system for inspection is the assembly of information that is needed for the risk-based approach. Some of this key assembly includes collecting engineering records, conducting field surveys, compiling experience from other sites and facilities, and interviewing cognizant personnel who operate and maintain the facility of interest. It is recommended that appropriate effort be placed on the interviewing of key personnel since they usually are knowledgable of degradation and errors that are not documented but are of utmost importance to the risk-based process.

In line with this human element, the next part of the process recommends the use of a qualitative risk assessment to capture fundamental expert judgment and experience in prioritizing systems, components, or elements for inspection. Bottom-up and top-down approaches are suggested to identify initiating failure events and consequences and to qualitatively estimate risk levels and ranking of components for inspection. Once again, a key element of the process is to thoroughly define failure modes and causes, including design, operational, and maintenance errors and a host of potential degradation mechanisms. This qualitative information facilitates the quantitative risk analysis.

The quantitative risk analysis is recommended to begin with an enhanced failure modes, effects, and criticality analysis (FMECA). The FMECA captures the information from the qualitative risk assessment and assigns probabilities of failure and consequences, in terms of safety and economic impact, for each component or element that is eligible for inspection. Operating experience databases and analytical models are recommended to be used to assist in the probability and consequence assignments. Once again, the elicitation of expert opinion from knowledgable personnel is a valuable source of information. The probability of failure for each element is multiplied by its respective consequence in order to obtain the risk. This process can be performed using single point estimates, or uncertainties in the probability and consequence values can be treated with interval estimates by defining lower and upper limits for these values. An interval analysis methodology is offered and recommended to perform the risk-based ranking. More complex uncertainty analysis methods are referenced. Both the single point estimate and interval analysis methods are exemplified for both single and multiple failure modes. Plots of the failure probabilities and consequences are suggested to perform the risk-based ranking. They include grouping components or elements based on failure probability only, consequence only, lines of constant risk, generalized risk contour lines, or special cases of the above (e.g., for insurance evaluations). Groupings or categories are suggested in order to facilitate the establishment of inspection programs for these components or elements.

A three-step process is recommended for the next part of the engineering risk analysis to develop the inspection program for each group or category of components or elements. This process is:

(1) choose candidate inspection strategies;
(2) choose an inspection strategy and perform the inspection; and
(3) choose appropriate action and update the state-of-knowledge.

Decision analysis logic is recommended in order to understand the approach. The choice of candidate inspection strategies must consider whether the components belong to an individual or set of facilities, the potential for damage actually exists, the potential for inspection damage will occur, and the reliability of the inspection method, including the potential for false calls and incorrect sizing if damage is found. In choosing an inspection strategy from the available candidates, structural reliability and risk assessment (SRRA) models are recommended to be exercised in order to determine the effect of inspection reliability, the potential for degradation mechanisms to exist, and the potential loadings on the failure probability of the component. Once a strategy has been selected and the inspection is performed, the SRRA models are exercised once again to choose appropriate action and to update the state-of-knowledge. Purely random inspections will have no real effect on the probability of failure of any facility or system. While the effect can be increased by concentrating the inspections into key areas, the primary value in a sample inspection rests in its ability to increase one's confidence in the safety of that facility or system. A methodology for determining this effect is offered. The SRRA models should also be exercised to evaluate corrective actions, relative to timing, the potential for success, and the potential for damage to be introduced, if significant findings occur. Regardless of what path is followed, these results are used to update the FMECA and the inspection strategy, and for some cases, redefinition of the system (i.e., Part 1) may be required.

A partial list of available software tools for performing parts of the risk-based inspection process is provided. However, it takes many tools to perform the entire process, and tools will have to be developed to meet this end. Some of the key tools that require further development include: database development (with appropriate graphical displays), simplified decision analysis and SRRA models to evaluate inspection strategies, and models to readily integrate inspection results into the process for updating the state-of-knowledge. Research for advancing the integration of cost-benefit analysis techniques with the engineering risk-analysis methods also needs to be performed for optimization of inspection programs.

To support this development work, it is recommended that pilot studies be performed using the recommended risk-based inspection process to applications of interest to the engineering community. The first studies are being performed for light water reactor nuclear power plant components and will be reported in Volume 2, which is to be used in conjunction with this general document. Application to other industries, such as fossil fuel-fired power plant components, petroleum refinery processing and storage components, and others, are also expected to occur so that appropriate tools are developed to meet the needs of these industries.

The primary benefit of the risk-based inspection approach is the focusing of inspection efforts on systems and components associated with the highest risk. Additional significant benefits are the insights that are gained in working through these processes and the enhancement of communications among the many disciplines that are involved to maintain an adequate level of safety within the affected industries and for society in general.

REFERENCES 6

6.1 ☐ TEXT REFERENCES

American Society of Mechanical Engineers, 1991 Draft (to be published), *Risk-Based Inspection – Development of Guidelines, Volume 2, Part 1, Light Water Reactor (LWR) Nuclear Power Plant Components,* ASME Research Task Force on Risk-Based Inspection Guidelines, Washington, D.C.

American Society of Mechanical Engineers, July 1990 (to be published), "White Paper on Reactor Vessel Integrity Requirements for Level A and B Conditions," ASME Boiler and Pressure Vessel Code, Section XI Task Group on Reactor Vessel Integrity Requirements, Draft Revision 7.

Balkey, K. R., Colburn, D. J., Meyer, T. A., and Paddleford, D. F., "Probabilistically-Based Inspection Guidelines," *Engineering Applications of Risk Analysis,* The American Society of Mechanical Engineers, December 1986, pp. 19–25.

Balkey, K. R., Meyer, T. A., and Witt, F. J., "Probabilistic Structural Mechanics: Chances Are. . .," *Mechanical Engineering,* Vol. 108, No. 7, July 1986, pp. 56–63.

Blanchard, B. S., Fabrycky, W. J., *Systems Engineering and Analysis,* Englewood Cliffs, N.J.: Prentice Hall, 1981.

Brown, R., Kahr, A., and Peterson, C., *Decision Analysis for the Manager,* New York: Holt, Rinehart, and Winston, 1974.

Bush, S. H., "Statistics of Pressure Vessel and Piping Failures," *Journal of Pressure Vessel Technology,* Vol. 110, August 1988, pp. 225–233.

Chapman, O. J. V., "A Statistical Approach to the Analysis of ISI Data Using the Bayes Method," Paper D1/7, 7th SMiRT Conference, Chicago, Illinois, 1983.

Chapman, O. J. V., "Confidence Through Sample ISI," 8th SMiRT Conference, 1985.

Chapman, O. J. V., "Probabilistic Based ISI and Life Extension," Paper M05/3, 10th SMiRT Conference, California, 1989.

Clevenger, T. G., "Organizing a Vessel, Tank, and Piping Inspection Program," ASME NDE-Vol. 5, *Nondestructive Evaluation: NDE Planning and Application,* pp. 29–34.

Dana, R., Sharp, D. E., and Webb, B. L., "Development of a Methodology for the Assessment of Residual Life of Pressure Vessels at Government Facilities," ASME NDE-Vol. 5, *Nondestructive Evaluation: NDE Planning and Application,* pp. 35–44.

Deluca, A., and Termini, S., "A Definition of a Nonprobabilistic Entropy in Setting of Fuzzy Sets Theory," *Information and Control,* Vol. 20, 1972, pp. 301–312.

Deluca, A., and Termini, S., "Entropy of L-Fuzzy Sets," *Information and Control,* Vol. 24, 1974, pp. 55–73.

Deluca, A., and Termini, S., "On the Convergence of Entropy Measures of Fuzzy Set," *Kybernetes,* Vol. 6, 1977, pp. 219–227.

Doctor, S. R., et al., "Reactor Safety Research Programs," *Quarterly Report July–September 1982,* NUREG/CR-2716, Pacific Northwest Laboratory, Richland, Washington, 1983.

Doctor, S. R., et al., "Non-destructive Examination (NDE) Reliability for Inservice Inspection of Light Water Reactors," *Semi-Annual Report April–September 1987,* NUREG/CR-4469, Vol. 7, November 1988.

Doctor, S. R., et al., "NDE Reliability and SAFT-UT Final Development," *Nuclear Engineering and Design,* Vol. 118, 1990, pp. 359–374.

Harris, D. O., Lim, E. Y., and Dedhia, D. D., "Probability of Pipe Fracture in the Primary Coolant Loop of a PWR Plant," Volume 5, *Probabilistic Fracture Mechanics Analysis,* NUREG/CR-2189, 1981.

Hartley, R. V. L., "Transmission of Information," *The Bell Systems Technical Journal,* Vol. 7, 1928, pp. 535–563.

Higashi, M., and Klir, G, J., "Measures of Uncertainty and Information Based on Possibility Distributions," *International Journal of General Systems,* Vol. 9, 1983, pp. 43–58.

Hohle, U., "Fuzzy Plausibility Measures," International Seminar on Fuzzy Set Theory, E. P. Klement, editor, Johanes Kepler University, Linz, Austria, pp. 7–30.

IEEE Standard 805, *IEEE Recommended Practice for System Identification in Nuclear Power Units and Related Facilities,* Institute of Electrical and Electronics Engineers, 1984.

Lercari, F. A., "Guidance for the Preparation of a Risk Management and Prevention Program," California Office of Emergency Services, November 1989.

Moghissi, A. A., et al., "Risk Assessment," *Mechanical Engineering,* Vol. 106, No. 11, November 1984, pp. 21–59.

Moore, R. E., *Interval Analysis,* Englewood Cliffs, N.J.: Prentice Hall, 1966.

Mosleh, A., Bier, V. M., and Apostolakis, A., "A Critique of Current Practice for the Use of Expert Opinions in Probabilistic Risk Assessment," *Reliability Engineering and System Safety,* Vol. 20, 1988, pp. 63–85.

Packman, P. F., et al, "The Applicability of a Fracture Mechanics Nondestructive Testing Design Criterion," *U.S. Air Force Report AFML-TR-68-32,* Air Force Materials Laboratory, Wright-Patterson AFB, Ohio, May 1968.

Perdue, R. K., "A Decision-Analytic Perspective for Corporate Environmental Risk Management," Westinghouse R&D Paper 89-1522-MABUS-P1, Unrestricted, Presented to the World Bank Workshop on Safety Control Risk Management, Washington, D.C., October 19, 1988.

Programme of Inspection for Steel Components, "Evaluation of the PISC-II Trial Results," PISC-II Final Report No. 5, Ispra, Italy, ISPRA Joint Research Center, September 1986.

Shannon, C. E., "The Mathematical Theory of Communication," *The Bell Systems Technical Journal,* Vol. 27, 1948, pp. 379–423 and 623–656.

Simonen, E. P., Johnson, K. I., Simonen, F. A., and Liebetrau, A. M., "VISA-II Sensitivity Study of Code Calculations," NUREG/CR-4614, November 1986.

Simonen, F. A., and Woo, H. H., "Analyses of the Impact of Inservice Inspection Using a Piping Reliability Model," NUREG/CR-3869, 1984.

Simonen, F. A., Johnson, K. I., Liebetrau, A. M., Engel, D. W., and Simonen, E. P., "VISA-II – A Computer Code for Predicting the Probability of Reactor Pressure Vessel Failure," NUREG/CR-4486, March 1986.

Spetzler, C. S., and Von Holstein, S. S., "Probability Encoding in Decision Analysis," *Management Science,* Vol. 22, 1975, pp. 340–358.

Taylor, T. T., Crowell, S. L., Doctor, S. R., and Posakony, G. J., "Detection of Small-Sized Near-Surface Under-Clad Cracks for Reactor Pressure Vessels," NUREG/CR-2878, 1983.

Technical Guidance for Hazard Analysis, U.S. Environmental Protection Agency, Washington, D.C., December 1987.

U.S. Nuclear Regulatory Commission, "Probabilistic Risk Assessment (PRA) Reference Document," NUREG-1050, September 1984.

U.S. Nuclear Regulatory Commission, "Reactor Safety Study: An Assessment of Accident Risks in U.S. Commercial Nuclear Power Plants," WASH-1400 (NUREG-75/014), October 1975.

Vesely, W. E., Davis, T. C., Denning, R. S., and Saltos, N., "Measures of Risk Importance and Their Applications," NUREG/CR-3385, Battelle Columbus Laboratories, Columbus, Ohio, 1983.

Von Winderfeldt, D., and Edwards, W., *Decision Analysis and Behavioral Research,* Cambridge University Press, 1986.

Yager, R. R., "Entropy and Specificity in a Mathematical Theory of Evidence," *International Journal of General Systems,* Vol. 9, 1983, pp. 249–260.

6.2 □ APPENDIX REFERENCES

American Association of State Highway and Transportation Officials, *AASHTO Maintenance Manual,* Washington, D.C., p. 263.

American Society of Mechanical Engineers, *Section III – Nuclear Power Plant Components, ASME Boiler and Pressure Vessel Code,* 1963, New York.

American Society of Mechanical Engineers, *Section XI – Rules for Inservice Inspection of Nuclear Power Plant Components, ASME Boiler and Pressure Vessel Code,* 1989, New York.

Ayyub, B. M., and Haldar, A., "Practical Structural Reliability Techniques," *Journal of Structural Engineering,* Vol. 110, No. 8, 1984, pp. 1707–1724.

Ayyub, B. M., White, G. J., and Purcell, E. S., "Estimation of Structural Service Life of Ships," *Naval Engineers Journal,* Vol. 101, No. 3, 1989, pp. 156–166.

Bamford, W. H., Thurman, A., and Mahlab, M, "Fatigue Crack Growth in Pressurized Water Reactor Feedwater Lines," Paper presented at the Joint Conference of the Pressure Vessels and Piping Materials, Nuclear Engineering and Solar Division, June 21–25, 1980, Denver, Colorado.

Berger, R. H., and Gordon, S., "Extending the Service Life of Existing Bridges," Transportation Research Board, No. 664, 1978, pp. 47–55.

Bush, S. H. and Maccary, R. R., "Development of Inservice Inspection Safety Philosophy for U.S. Nuclear Power Plants," BNWL-SA-4129, Pacific Northwest Laboratory, Richland, Washington, 1972.

Carr, P., et al., "A Probabilistic Strategy for Subsea Inspection of Steel Structures," SPE European Petroleum Conference, London, 1986.

Chapman, O. J. V., "A Statistical Approach to the Analysis of ISI Data Using the Bayes Method," Paper D1/7, 7th SMiRT Conference, Chicago, Illinois, 1983.

Chapman, O. J. V., "Probabilistic Based ISI and Life Extension," Paper M05/3, 10th SMiRT Conference, California, 1989.

Christian, T. F., Smith, H. G., and Saff, C. R., "Structural Risk Assessment Using Damage Tolerance Analysis and Flight Usage Data," presented at American Society of Mechanical Engineers 1986 Winter Annual Meeting.

Cottrell, B. H., Jr., "Field Evaluation of Highway Safety Hardware Maintenance Guidelines," Transportation Research Board, No. 1163, 1988, pp. 63–69.

Crist, M., "Inspection Frequency Criteria Models for Timber, Steel, and Concrete Pile Supported Waterfront Structures," Naval Civil Engineering Laboratory, No. NCEL-CR-87.005, Port Hueneme, California, 1986, p. 132.

Doctor, S. R., et al., "Nondestructive Examination (NDE) Reliability for Inservice Inspection of Light Water Reactors," *Semi-Annual Report, April–September 1988,* prepared for U.S. Nuclear Regulatory Commission by Pacific Northwest Laboratory, Richland, Washington, NUREG/CR-4469, Vol. 9, 1989.

Fujimoto, Y., Itagaki, H., Itoh, S., Asada, H., and Shinozoka, M., "Bayesian Reliability Analysis of Structures With Multiple Components," *Structural Safety and Re-*

liability: Proceedings of ICOSSAR 89, the 5th International Conference on Structural Safety and Reliability, Vol. 3, 1989 pp. 2143–2146.

Gore, B. F., and Huenefeld, J. C., "Methodology and Application of Surrogate Plant PRA Analysis to the Rancho Seco Power Plant, Volumes 1 and 2," NUREG/CR-4768, PNL-6032, U.S. Nuclear Regulatory Commission Region V, Walnut Creek, California, July 1987.

Gore, B. F., Vo, T. V., and Harris, M. S., "PRA Applications Program for Inspection of Oconee Unit 3," NUREG/CR-5006, U.S. Nuclear Regulatory Commission Region I, King of Prussia, Pennsylvania, October 1987.

Gore, B. F., Vo, T. V., Colburn, A. J., Eschbach, E. J., Harris, M. S., and Throm, E. D., "Low Temperature Brittle Fracture Risk Reduction for PWR Reactor Vessels," *Nuclear Safety,* Vol. 31, No. 1, January–March 1990.

Harrison, J. W., and Grenke, W. C., "Procedure for Determining Frequencies to Inspect and Repair Highway Safety Hardware," FHWA Report No. FHWA-IP-83-4, 1983, p. 25.

Hazelton, W. S., and Koo, W. H, "Technical Report on Material Selection and Processing Guidelines for BWR Coolant Pressure Boundary Piping – Final Report," NUREG-0313, Rev. 2, U.S. Nuclear Regulatory Commission, Washington, D.C., 1988.

Heaberlin, S. W., et al., "A Handbook for Value-Impact Assessment," NUREG/CR-3568, prepared for U.S. Nuclear Regulatory Commission by Pacific Northwest Laboratory, Richland, Washington, 1983.

Imbsen, R. A., and Schamber, R. A., "Proposed Limit State Strength Evaluation of Existing Reinforced-Concrete Bridges," Transportation ResearchBoard, Vol. 2, No. 950, 1984, pp. 92–100.

Karydas, D. M., "Probabilistic Analysis for Low Water Level Conditions in Boilers," The American Society of Mechanical Engineers, ASME Paper 87-WA/SAF-9, New York, 1987.

Kesling, G. D., and Whittaker, I. C., "Simulation Model for Railroad Reliability," *Simulation,* Vol. 44, No. 4, 1985, pp. 168–180.

Kitagawa, Hideo, "Some Recent Life Prediction Methods Based on Material Technology and Structural Mechanics," *Structural Safety and Reliability: Proceedings of ICOSSAR 85, the 4th International Conference on Structural Safety and Reliability,* Vol. 3, 1985, pp. 429–438.

Knepp, A. J., Tseng, M. T., and Jones, J. S., "Benefit-Risk Analysis in the Design of Highway Stream Crossing," *Public Roads,* Vol. 40, No. 2, 1976, Monograph, pp. 66–90.

Madsen, H. O., Skjong, R. K., Tallin, A. G., and Kirkemo, F., "Probabilistic Fatigue Crack Growth Analysis of Offshore Structures With Reliability Updating Through Inspection," *Marine Structural Reliability Symposium,* The Society of Naval Architects and Marine Engineers, Jersey City, N.J., 1987, pp. 45–55.

Mohammadi, Jamshid, and Longinow, Anatol, "Probability-Based Design of Highway Bridges," *Structural Safety and Reliability: Proceedings of ICOSSAR 85, the 4th International Conference on Structural Safety and Reliability,* Vol. 3, 1985, pp. 23–32.

Mohammadi, Jamshid, and Yazbeck, G. John, "Strategies for Bridge Inspection Using Probabilistic Models," *Structural Safety and Reliability: Proceedings of ICOSSAR 89, the 5th International Conference on Structural Safety and Reliability,* Vol. 3, 1989, pp. 2115–2122.

Moses, F., and Verma, D., "Load Capacity Evaluation of Existing Bridges," NCHRP Report N301, Transportation Research Board, 1987, p. 110.

National Transportation Safety Board Bureau of Accident Investigation, Washington, D.C., 1984, "Highway Accident Report – Collapse of a Suspended Span of Interstate Route 95 Highway Bridge Over the Mianus River, Greenwich, Connecticut, June 28, 1983," Report No. NTSB/HAR-84/03, p. 111.

National Transportation Safety Board Bureau of Accident Investigation, Washington, D.C., 1985, "Highway Accident Report – Collapse of the U.S. 43 Chickasawbogue Bridge Spans Near Mobile, Alabama, April 24, 1985" Report No. NTSB/HAR-86/01, p. 41.

National Transportation Safety Board Bureau of Accident Investigation, Washington, D.C., 1988, "Highway Accident Report – Collapse of New York Thruway (I-90) Bridge Over the Schoharie Creek, Near Amsterdam, New York, April 5, 1987," Report No. NTSB/HAR-88/02, 4656A, p. 168.

Nilsson, F., Skanberg, L., and Bystedt, P., "New Swedish Regulations for Safety of Pressurized Components," presented at the Ninth International Conference on Nondestructive Evaluation in the Nuclear Industry, Tokyo, Japan, April 25–28, 1988.

Parekh, I. R., Graber, D. R., and Berger, R. H., "A Comprehensive Bridge Posting Policy," Transportation Research Record No. 1083, 1986, pp. 35–45.

Peterson, K. E., "Risk Analysis Used as a Tool in Selection Between Alternative Traffic Systems," Proceedings of the 5th IFAC/IFIP/IFORS Conference on Control in Transportation Systems, IFAC, Austria, IFIP, Switzerland, 1986.

Program for the Inspection of Steel Components (PISC-II), "Evaluation of the PISC-II Trials Results," PISC-II Report No. 5, September 1986, Final Issue, CNSI No. 121, Organization for Economic Cooperation and Development, Nuclear Energy Agency, Committee on the Safety of Nuclear Installations, CNSI, No. 121.

Raczon, F., "A Guide for the Roads & Bridges Industry," Scranton Gillette Communication, Incorporated Roads & Bridges, Vol. 25, No. 12, 1987, pp. 49–61.

Saff, C. R., Smith, H. G., and Christian, T. F., "Application of Damage Tolerance Analysis to Structural Risk Assessment," 28th Structures, Structural Dynamics and Materials Conference, AIAA/ASME/ASCE/AHS, April 1987.

Shanafelt, Go, and Horm, W. B., "Guidelines for Evaluation and Repair of Damaged Steel Bridge Members," Transportation Research Board, Report No. N271, 1984.

Shirole, A. M., and Hill, J. J., "Systems Approach to Bridge Structure Replacement-Priority Planning," Transportation Research Board, No. 664, 1978 pp. 32–36.

Simonen, F. A., "An Evaluation of the Impact of Inservice Inspection on Stress Corrosion Cracking of BWR Piping," ASME PVP-Vol. 186/NDE-Vol. 7, *Codes and Standards and Applications for Design and Analysis of Pressure Vessel and Piping Components,* 1990, pp. 187–193.

Society of Automotive Engineers, "Fault/Failure Analysis Procedure," ARP 926A, November 1979.

Society of Automotive Engineers, "Fault/Failure Analysis for Digital Systems and Equipment," ARP 1834, August 1986.

Smith, H. G., Saff, C. R., and Christian, T. F., "Structural Risk Assessment and Aircraft Fleet Maintenance," 31st Structures, Structural Dynamics, and Materials Conference, AIAA/ASME/ASCE/AHS, April 1990.

Tallin, A., Kirkimo, F., and Madsen, H., "Probabilistic Fatigue Crack Growth Analysis With Reliability Updating," *Proceedings of the 5th ASCE Specialty Conference on Probabilistic Methods in Civil Engineering,* ASCE, 1988.

U.S. Department of Defense, "System Safety Program Requirements," Mil-Std-882B, March 1984,.

U.S. Nuclear Regulatory Commission, "Quality Group Classifications and Standards for Water, Steam, and Radioactive Waste Containing Components of Nuclear Power Plants," NRC Regulatory Guide 1.26 Revision 1, 1974.

U.S. Nuclear Regulatory Commission, "Regulatory Guide 1.154 – Format and Content of Plant-Specific Pressurized Thermal Shock Safety Analysis Reports for Pressurized Water Reactors," January 1987.

U.S. Nuclear Regulatory Commission, "Risk-Based Operational Safety and Performance Assessment," *Inspection Manual,* Chapter 2515: Operations, Procedure No. 93804, Washington, D.C., November 25, 1988.

U.S. Nuclear Regulatory Commission, "General Design Criteria for Nuclear Power Plants," 10 CFR 50, Code of Federal Regulations, Chapter 10, Part 50, Appendix A, November 30, 1988 (Latest Edition).

U.S. Nuclear Regulatory Commission, "Probabilistic Risk Assessment Technology Training Program," April 17–21, 1989, PRA Basics for Inspection Applications – Course Documentation, Washington, D.C.

Vo, T. V., Gore, B. F., and Harris, M. S., "Probabilistic Risk Assessment Based Inspection Guidance for Arkansas Nuclear One Unit 1," *Nuclear Technology,* Vol. 84, American Nuclear Society, LaGrange Park, Illinois, Janaury 1989.

White, G. J., and Ayyub, B. M., "Reliability Methods for Ship Structures," *Naval Engineers Journal,* ASNE, Vol. 97, No. 4, 1985, pp. 86–96.

White, G.J., and Ayyub, B. M., "Reliability-Based Fatigue Design of Ship Structures," *Naval Engineers Journal,* ASNE, Vol. 99, No. 3, May 1987.

Winget, R.R., "Point Beach Nuclear Plant Secondary In-Service Inspection Project and Badness Factor Program," *Nuclear Technology,* Vol. 85, April 1989, pp. 7–11.

Yazdani, N., and Albrecht, P., "Risk Analysis of Fatigue Failure of Highway Steel Bridges," *Journal of Structural Engineering,* ASCE, Vol. 113, No. 3, 1987, pp. 483–500.

APPENDIX A
SUMMARY OF CURRENT INSPECTION REQUIREMENTS AND RELATED DEVELOPMENTS IN GENERAL

This Appendix reviews risk-based methods and current inspection requirements as they are practiced in the various industries of interest to this document. Examples are cited that show increasing acceptance and use of risk-based approaches in the development of inspection programs. The discussion shows that past inspection practices have often made use of implied considerations of risk in a qualitative sense, in combination with engineering judgment. More recently, quantitative risk methods have begun to see selective applications to guide decision-making in the area of inspection planning. This information provided a starting point to build the recommended risk-based inspection process given in the main body of this general document.

NUCLEAR APPLICATIONS A1

A1.1 ☐ INSPECTION OF PRESSURE BOUNDARY COMPONENTS

The nuclear power industry has been the subject of strict regulatory requirements as well as intense concern for public safety. Accordingly, the inspection of structural systems and components has been part of the generally high level of oversight given to all aspects of the construction, maintenance, and operation of nuclear power plants. The discussion here specifically focuses on inservice inspections (ISI) of pressure vessels and piping systems. Such inspections are performed in accordance with the *Rules for Inservice Inspection of Nuclear Power Plant Components* as specified by Section XI of the ASME Boiler and Pressure Vessel Code (BPVC) (1989 Edition). These rules have been accepted by the U.S. Nuclear Regulatory Commission (NRC), and it is required that all operating nuclear power plants perform periodic inspections in accordance with ASME BPVC Section XI requirements, with these inspections supplemented by additional NRC-directed inspections to address specific concerns with service-induced structural degradation.

The philosophy and approach behind ASME BPVC Section XI has been documented by Bush and Maccary (1972). Prior to the first publication of ASME BPVC Section XI in 1970, there were no requirements for inservice inspection because:

- ISI was considered impractical due to radioactivity levels;
- ISI was considered unnecessary due to the enhanced criteria applied during design and construction [i.e., ASME BPVC Section III (1963)];
- No special provisions for accessibility were provided to enable ISI.

After publication of ASME BPVC Section III in 1963, the former Atomic Energy Commission (AEC) recognized in 1966 that enhanced quality standards did not justify omission of ISI. A joint AEC-industry effort subsequently produced a "Draft Code for Inservice Inspection of Nuclear Coolant Systems" in 1968. The initial draft code was basically limited to pressure vessels, but its overall scope was sufficient to accommodate its present broader coverage of nuclear power plant components. The major requirements and features of ASME BPVC Section XI, as first published in 1970, included:

- a recommendation for the future design of nuclear systems to accommodate ISI and possible repairs;
- pre-service (baseline) examination prior to startup;
- a wide range of prescribed nondestructive examination (NDE) techniques, (ranging from ultrasonic methods to visual examinations) based on the particular component to be inspected and the relative importance of the inspection to plant safety;
- acceptability of new NDE techniques for ISI (when validated);
- timing and extent of inspection for each component based on assumptions relative to the safety significance and consideration of expected degradation;
- inspections performed using representative, statistical sampling plans with major attention directed to welds;

* repetitive inspection intervals of 10 years, with the required inspections performed throughout this interval.

Legal and regulatory requirements dictate the inspections of nuclear facilities. Nuclear power plants are required by the NRC through Appendix A of 10 CFR 50 [1988 (latest edition)] to ensure that all structures, systems, and components important to safety be designed, fabricated, erected, and tested to quality standards commensurate with the importance of the safety functions to be performed. In the specific case of the reactor coolant system (RCS) pressure boundary, Section 10 CFR 50.55a states that ASME requirements for Class 1 components are applicable. This is a direct reference to the ASME BPVC approach of defining the most stringent requirements for Class 1 components, and less stringent requirements for Class 2 and 3 components, respectively.

For Class 2 systems, U.S. NRC Regulatory Guide 1.26 (1974) acknowledges 10 CFR 50.55a, and elaborates on it by requiring that ASME rules for Class 2 components be met by certain systems or portions of systems important to safety that are designed for:

* emergency core cooling
* post-accident heat removal
* post-accident fission product removal
* reactor shutdown
* residual heat removal
* steam and feedwater systems

Similarly, ASME Class 3 requirements must be met by other specified systems or portions of systems that have a lower level of safety significance.

Although the development of the present ISI standards used in the nuclear industry did not include formal considerations of risk, it is apparent that perceived risk was considered, since the most thorough inspections are reserved for those systems and components for which the consequences of failure are the greatest. Similarly, while ASME BPVC Section XI requires that only about 25% of the welds be sampled in the ISI program, these requirements specify that this sample include those welds which are believed to be most likely to fail (e.g., high-static stresses, high-fatigue usage, dissimilar metal welds, terminal ends of piping runs, etc.). A sequential expansion of the sample is required in those cases that the inspections find indications of flaws during the 25% sample.

In describing inspection efforts for nuclear power plants, ASME BPVC Section XI has provided a baseline of minimum requirements for periodic inspection. These inspections have greatly enhanced the general level of confidence in the soundness of the original designs and fabrication quality. In some cases, plant operating experience has led the U.S. NRC to require additional inspections beyond minimum code requirements, mainly for specific systems and components where the original designs have proven to be inadequate due to unanticipated stresses and material degradation.

At some plants, cracking, e.g., thermal fatigue, as presented by Bamford, et al. (1980), has occurred at locations where no ASME BPVC Section XI inspection was required. At other plants, cracking, e.g., stress corrosion cracking in stainless steel piping, as reported by Hazelton and Koo (1988), has occurred at inspected welds, but the code inspection methods (ultrasonics) proved to be ineffective in detecting this cracking. The nuclear industry has responded to regulatory authorities and devoted considerable effort and expense both to performing extra inspections and to upgrading inspection reliability. For some components, inspection sample sizes and inspection frequencies have also been dramatically increased in response to certain recurring service issues. For example, the eddy current inspection of steam generator tubes responded to issues associated with the tubes. Formal changes to inspection requirements have been implemented into ASME BPVC Section XI, often rather quickly, to respond to needs and to correct known deficiencies. However, a systematic review and updating of code inspection plans has not occurred as industry knowledge and operating experience has increased over the years.

While there has been a growth in ISI requirements over the years, it should also be noted that certain code-mandated inspections have been deleted from inspection plans with the approval of regulatory authorities (relief requests), when such inspections have proven to be impractical. Lack of physical access to welds and lack of suitable techniques to detect defects have been the reasons for deleting the inspections at locations which otherwise have had favorable service experience with no evidence of degradation. In future years, relief requests may become more difficult to justify, as concerns with the structural integrity of older plants must be addressed.

The philosophy and assumptions originally used in developing ASME BPVC Section XI have generally proven valid. However, major advances have since been made in the available methods to quantify the underlying risk-based concepts of consequence and probability. Specifically, the nuclear industry is a leading user of probabilistic risk assessment to calculate risks associated with accidents. Similarly, probabilities of structural failures can now be better estimated from the growing database from many years of plant operating experience. There are also computer codes (using probabilistic fracture mechanics) for estimating probabilities for the types of catastrophic structural failures that are too low to estimate from historical operating experience.

There is a growing interest in the nuclear industry to review the assumptions and detailed requirements of ASME BPVC Section XI. The formation of the ASME Research Task Force charged with writing this report is a direct outcome of this interest. Along similar lines, the Electric Power Research Institute and the U.S. Nuclear Regulatory Commission have each funded a research project directed at developing improved inspection requirements. The Pressure Vessel Research Committee is also performing an in-depth review of both ASME BPVC Sections III and XI with an emphasis on simplification and clarification of requirements.

The accumulated results of ISI findings and the lack of severe accidents from vessel rupture and pipe breaks tends to confirm PRA studies, which conclude that potential structural integrity failures are a relatively small contribution to overall plant risk. Functional failures of active components (e.g., valves) give greater contributions. Such failures are outside the scope of this report. The fact that failures have occurred at locations not inspected by current ISI plans has suggested that current ISI requirements for pressure boundary integrity should be reexamined. The goal is to optimize inspections with decreased inspections where the consequences and/or the estimated probabilities of failure are small, and increased inspections for higher priority systems and components.

In general, the nuclear industry has experienced a rapid growth in the quantified applications of risk-based methods as a guide to improved safety requirements. A notable application has been the use of PRA as an aid to the NRC personnel who serve as on-site inspectors at nuclear power plants, as discussed in the next section. In "walkdown" type inspections, limited time and personnel must be carefully allocated to check items of the highest priority to the safe operation of the plant. NRC has been using PRA-based information to develop inspection guidance, as shown by Vo, et al. (1989).

In contexts other than inspection, there are extensive applications of risk-based methods in the nuclear industry. For example, the issue of the possible fracture of reactor vessels during pressurized thermal shock events was addressed using PRA methods to estimate event frequencies and severity, as evidenced in U.S. NRC Regulatory Guide 1.154 (1987). In such studies, probabilistic fracture mechanics has been applied to estimate vessel and piping failure probabilities for different event scenarios. Also, NRC routinely applies risk-based methods to review the appropriateness of potential changes in the regulations to be imposed on the nuclear power industry, by performing a "value impact analysis" which addresses cost/benefit considerations of risk reduction alternatives. Heaberlin, et al. (1983) and Gore, et al. (1990) provide a handbook and application of these assessment methods, respectively.

A1.2 □ U.S. NUCLEAR REGULATORY COMMISSION RISK-BASED INSPECTION PROGRAM

A program is currently being implemented for inspections dealing with operational safety verification, maintenance, and surveillance requirements for nuclear power plants. Although this program can enhance the inspections that are performed to find flaws and evaluate degradation mechanisms for insuring the integrity of pressure boundary components, these inspections primarily address active components, which are outside the scope of this general document. The objective of the program, as stated in the current NRC Five Year Plan, is "to assess licensees' operation of nuclear power plants to ensure safe operation of the facilities in accordance with NRC regulations." An underlying objective supporting this overall objective is "to ensure that the finite resources available for inspections are efficiently and effectively allocated to enhance reactor safety." A key element in this program is the use of insights from probabilistic risk assessments (PRA) to focus inspection activities on the most risk-significant areas and issues.

Risk insights obtained from PRAs are incorporated into this program in two major ways. First, plant-specific "Risk-Based Inspection Guides" are being prepared which identify the most risk-important systems and components at each plant. This information is provided to NRC resident inspectors for use in planning their routine plant inspection activities. Second, risk information from PRAs is being used directly in the planning and performance of team inspections by NRC regional and headquarters inspectors. Additional analysis of PRA information is required for both of these uses. In support of these activities, the NRC is providing additional training to inspection personnel in understanding and applying the information which is provided in PRAs.

NRC Course "PRA Basics for Inspection Application." To support the risk-based inspection program, a U.S. NRC PRA Training Program (1989), which lasts five days, has been developed that introduces NRC resident inspectors to event trees, fault trees, and the calculation and interpretation of risk importance measures such as the Fussel-Vesely and Birnbaum Importance Measures. It also introduces them to the extensive qualitative information on safety systems, which is presented in PRAs, including the systems descriptions, dependency diagrams, success criteria, common cause failure relationships, and dominant event sequences. Major insights obtained from PRA analyses performed to date are also presented.

This course provides the inspectors with the ability to understand and selectively extract information from PRAs, without being overwhelmed by the multi-volume reports or intimidated by the terminology. It also introduces the inspectors to the Risk-Based Inspection Guides and the method of analyzing PRA information used to develop inspection plans for risk-based team inspections.

Risk-Based Inspection Guides. The Risk-Based Inspection Guides, as developed by Gore, et al. (1987), present risk information derived from plant PRAs in a format that does not require PRA knowledge to understand or apply. The format has been developed to be immediately useful to NRC resident inspectors in planning and performing their routine inspection activities, and evaluating the risk significance of inspection findings and operational events. Guides have been, or are being, produced for all plants for which PRAs are available to the NRC.

Each guide presents a discussion of the most risk-important accident initiators for the plant, followed by descriptions of the dominant accident sequences that lead to core damage. When possible, plant design features that cause unusual vulnerabilities to specific accident initiators are identified.

This discussion is followed by a listing of plant systems associated with 98% of the inspectable risk of core damage. This list is ordered in two ways and according to two different risk-importance measures. These are the fraction of the total core damage frequency, to which failures of components in each system contribute (the Fussel-Vesely Importance Measure), and the probability of core damage assuming that the system has failed (the Birnbaum Importance Measure). These prioritization schemes are based on the failure probability estimates used in the PRA analysis. Inspections are then performed to prevent degradation of these probabilities (current PRAs do not consider long-term degradation).

For each of these risk-important systems, a brief system description is provided, along with other important information provided in, or developed from, the PRA. For each system, risk-important components are tabulated and inherent major failure modes are described. Single failures and unusual system vulnerabilities identified from the PRA are also noted.

In addition, an abbreviated system walkdown checklist is provided addressing only the risk-important components. This table allows the resident to perform relatively rapid system inspections which address most of the risk associated with the system.

Cross-reference tables are also provided which list the component failure modes relevant to specific types of inspection efforts such as maintenance, surveillance, and operator actions.

Since not all plants have PRAs available for analysis, a technique has been developed to produce inspection guides for plants lacking PRAs. It uses insights from published PRAs regarding the risk importance of safety and support functions to infer the risk important systems and components at other plants, as developed by Gore and Huenefeld (1987). Test comparisons between guides produced this way, and guides produced by analyzing the PRAs directly, indicate that this procedure produces meaningful and useful risk-based inspection guidance when PRA information is not available.

Risk-Based Operational Safety and Performance Assessment. This U.S. NRC Inspection Manual (1988) utilizes risk information presented in a power plant PRA to structure and prioritize the on-site inspection activities of a team of technical experts engaged in an extended inspection of one to several weeks duration. A methodology is provided for the identification (from the PRA) of risk-important components, and of important accident mitigation and recovery actions, for use in assessing the operational readiness of the power plant.

This inspection procedure focuses on determining whether or not the plant is operated and maintained in such a way that:

- component failures, which may challenge plant safety systems, are minimized;
- safety systems, equipment, and components will be available, reliable, and operable;

- plant operators are capable of recognizing and responding appropriately to plant challenges, and capable of conducting timely and effective accident mitigation and recovery actions;
- the licensee has appropriately factored available risk information into the plant's programs, procedures, and design.

The procedure describes how PRA information should be used to rank order the risk importance of component and operator failures (basic events), based on the core damage frequencies associated with the most important component failure combinations. Initiating events, component and instrumentation failures, and operator errors and recovery actions are all addressed. Based on this information, the lines of inquiry of the inspection and the composition of the inspection team are to be driven by the PRA information. An important aspect of this inspection procedure is that it addresses operator performance in recognizing and responding to off-normal situations, and in mitigating the effects of accidents. Risk-important accident sequences from the PRA are used to develop simulator scenarios, which are then used to test the responses of plant operators and the adequacy of the plant procedures for coping with the accidents.

A1.3 □ RISK-BASED INSPECTION IN THE UNITED KINGDOM

One of the most active areas in the United Kingdom with regard to inservice inspection and safety assessment is the extended life justification of the Magnox generation of nuclear installations. The basis for extended life of these installations is a long-term safety review. As part of this review, an extensive amount of ultrasonic inspection has been carried out mainly on the gas ducting but also on a limited amount of thicker boiler shells. The philosophy of this inspection has, however, not been based on formal probabilistic analysis. The philosophy has been first to carry out a sample inspection of all the different features, concentrating particularly on welds. Following this, if defects are detected, all similar features are inspected; otherwise it may be judged that no further inspection is required, at least in the short term. Although no formal probabilistic method has been applied to this inspection, it is apparent that there are inherent probabilistic assumptions in the procedure. The combination of failure probability and consequence (i.e., risk), however, is not considered.

A second area of nuclear-related activity is in the United Kingdom Nuclear Submarine program. Chapman (1983 and 1989) shows a probabilistic-based approach for both optimizing and measuring the gain in confidence from inservice inspection of vessel and piping components. The model uses an expert system together with mathematical modeling to form an initial best estimate of the start of life defect distribution for welds. The through-life history is then calculated to arrive at an end of life failure probability. A series of inspection programs can now be applied through life and their effect on the failure probability calculated. Clearly, the results, and hence conclusions, about the optimum inspection depend on the initial assumptions and judgments. In order to overcome this, once an inspection program is set out and results become available, Bayesian logic is applied to gain confidence in the initial defect distribution and through-life prediction.

Moving away from the nuclear field, there appears to be little movement toward a risk-based inspection logic except in the inspection of oil rigs in the North Sea. Here, a method by Carr (1986), which is very similar in principle to that discussed for the Nuclear Submarine program, has been used. Crack initiation followed by crack growth is the assumed failure mechanism, meaning that no initial defect distribution for the joints is required. Having built the probabilistic model, a series of inspection programs can be simulated to arrive at an optimum program. As with the previous case, the optimum program is a function of the basic assumptions and so, again, Bayesian logic is used in a feedback loop to gain confidence in these assumptions.

In both of the above applications of probabilistic-based inspection, the driving force to optimize is the need to minimize the risk to human operators, be it from the obvious dangers of the North Sea or the less tangible dangers of irradiation. However, since the objective is to obtain the least risk for a given effort, it is logical to extend this optimization to a financial consideration.

A1.4 □ NDE RELIABILITY

The detection of defects by nondestructive examination (NDE) is often a difficult task, particularly if safety requirements dictate a high level for the "probability of detection" for crack-like defects. To quantify the reliability of current and improved NDE methods, the U.S. NRC has supported a long-term research program at PNL, as described by Doctor (1989). The European community has addressed similar objectives over the last decade through the Programme for Inspection of Steel Components (PISC)-II and PISC-III research programs (1986). Research results have revealed a number of serious shortcomings in industry practices for field inspections, and more importantly the nuclear power industry has benefited from resulting upgrades to ASME BPVC Section XI.

Figure A-1 shows how enhanced NDE reliability can improve the safety and reliability of reactor piping systems, as reported by F. A. Simonen (1990). Fracture mechanics calculations have predicted rapidly growing stress corrosion cracks in the stainless steel pipes of boiling water reactors. Timely detection and repair of these cracks can lead to marked enhancements in system reliability. The performance of field inspection teams was first carefully measured in a controlled testing environment. Performance data for the top teams ("good" NDE reliability) was statistically quantified and was proven to be clearly superior to the corresponding performance of the least qualified teams ("poor" NDE reliability). As indicated in Fig. A-1, the "good" NDE performance can improve the piping reliability by nearly a factor of 10.

Furthermore, the expected gain in performance from new and improved technology ("advanced" NDE reliability) can provide another factor of 10 improvement in piping reliability. Since the "advanced" inspections can also be performed less frequently, improved technology is a winning proposition for cost-benefit reasons, mainly because costly reactor downtime is significantly reduced.

Quantitative risk techniques are also currently being developed in other countries, such as Sweden and Germany, in support of NDE requirements for reactor system components. For example, in Sweden the inspection of structural systems and components is guided by a recently adopted methodology that systematically assigns inspection requirements on the basis of consequences and probabilities, as reported by Nilsson, et al. (1988). While the Swedish approach is qualitative, quantitative results can be used to guide the assignment of components to a relative scale of consequence and risk categories. In a sense, the Swedish approach is similar to that implied by ASME BPVC Section XI, but goes one step further by requiring each nuclear plant to develop an individual plan based on an ongoing review of updated knowledge of plant design features and plant operating experience.

For secondary piping systems, Point Beach Nuclear Plant personnel have developed an inservice inspection program to detect and quantify significant service-related degradation and preexisting conditions that could jeopardize the integrity of those systems in the future. A "Badness Factor Program" has been developed by Winget

(1989) to rank components as to their susceptibility to erosion-corrosion and stress-induced fatigue. It uses hydrodynamic variables to assign a factor to each component and pipe fitting so that a comparison of the relative magnitudes of this factor can be made between systems or piping sections. Although a qualitative decision is then made to identify locations for degradation, the approach provides useful insights for development of a quantitative risk-based approach to optimize the inspection of the thousands of locations within these systems.

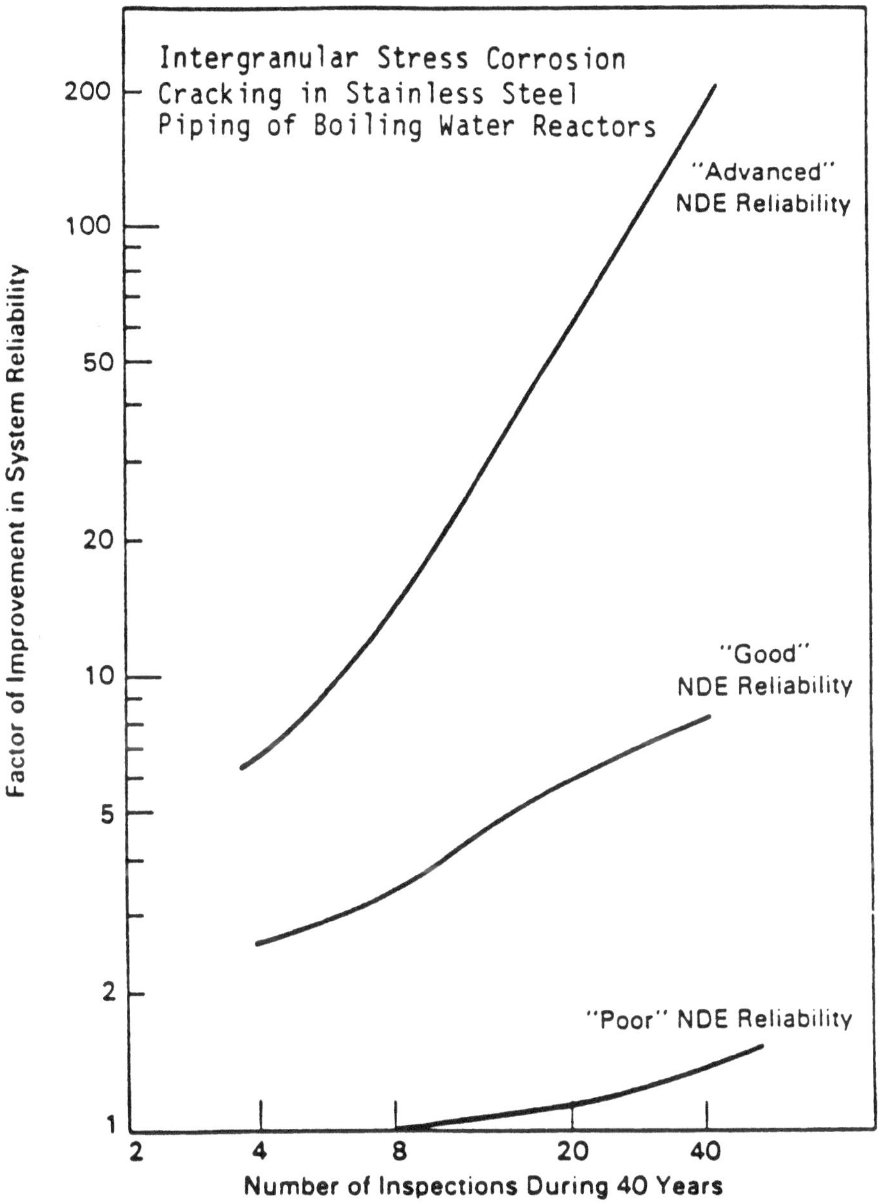

FIG. A1 ENHANCEMENT OF PIPING RELIABILITY BY IMPROVEMENTS IN NDE [Simonen, F. A. (1990)]

FOSSIL FUEL-FIRED POWER PLANT COMPONENTS

A2

The potential off-site consequences of component failures in fossil fuel-fired power plants are much less severe than those for nuclear power plants or even some petroleum and chemical processing plants. Consequently, a higher level of failure has been tolerated. This, along with the large population of fossil plants and the long time period over which such plants have operated, has led to a relatively large reliability experience base in fossil plant components. Use of this experience is, in fact, probabilistic-based inspection planning, where extensive failure data is available to determine failure probabilities. Consequently, inservice inspection of fossil plant components (where and when to inspect) is largely based on experience, including experience with a given plant, within a given utility, or among groups of utilities. Information on failures is shared among utilities, through channels such as the Edison Electric Institute or the Electric Power Research Institute. News of dramatic and unusual failures travels quickly through the utility industry and often leads to a rapid concerted inspection and mitigation effort to avert similar failures at other plants. The rapid response to high-consequence failures likewise indicates an awareness of risk, although it is not formally quantified.

Safety issues involved in fossil plant component failures have been relatively minor in terms of public risk. Therefore, inspections are more focused toward reducing risk to on-site personnel and minimizing costs. Forced outages of a plant are much more expensive than planned outages, which can be scheduled at times of reduced power demands, such as during the spring and fall. Component repair or replacement occurs routinely during a scheduled outage and is usually less expensive during a scheduled outage than during a forced outage. Consequently, inservice inspections of fossil plant components are aimed to a large extent at averting failures that lead to forced outages.

Some components in fossil plants have useful lifetimes that are much less than the useful lifetime of the plant and are replaced routinely, preferably during a scheduled outage. Examples of this are water-wall tubes and superheater/reheater tubes. Failure of some such tubes during service can be tolerated, with repair being performed during the next scheduled outage. Inspections of such tubes are often performed, with the frequency and location of inspections based on past experience. The results of such inspections assist in planning of repairs during the current and future outages.

Some components are inspected during planned outages to detect potential issues that have been identified in other plants. Examples of these are thick-walled boiler components (headers) and seam-welded reheat lines. Such inspections are aimed at minimizing the likelihood and economic consequences of a forced outage, as well as the potentially severe on-site consequences of a failure. The results of such inspections are intended to foresee a coming issue and to allow for timely acquisition of replacement parts.

Still other components are occasionally inspected during planned outages because of generic potential issues with that component. Prime examples are the rotor, disks, and blades of the steam turbine, and the rotor and retaining rings of the generator. Inspections of turbine components, especially the disks and rotors, are performed to identify potential issues of material degradation – especially cracking. Early detection of issues allows repair or replacement in a timely manner, thereby minimizing forced outages, as well as the potentially severe on-site consequences of a rotor failure. The lead time for acquisition of replacement parts can be years, so early warning of an impending replacement is needed. Also, the cost of repair may be much greater if damage is allowed to progress.

The results of inspections also provide valuable inputs to run/retire decisions. Such decisions are becoming more common in conjunction with life-extension efforts for installed fossil fuel-fired generating capacity. Increased emphasis is being placed on periodic inspections as older plants are continued in service.

AIRCRAFT STRUCTURES A3

Risk assessment plays a vital role in the aircraft industry, both commercial and military. Safety, fleet management, and life extension all require the use of risk-assessment methods. The methods used fall into two basic categories:

- probabilistic risk assessment
- probabilistic fracture mechanics

A3.1 □ PROBABILISTIC RISK ASSESSMENT

Commercial Aircraft. Federal Aviation Regulation Part 25 is the governing document for the commercial aircraft industry. It specifies that the analysis be done, but does not say how it should be conducted. Recommended procedures for risk analysis have been prepared by the Society of Automotive Engineers in ARP 926A (1979) and ARP 1834 (1986), which are used for commercial aircraft applications. The procedures describe a progression of analysis methods beginning with a traditional failure mode, effects, and criticality analysis (FMECA) and progressing to fault tree analysis. The FMECA analysis can be performed either as a hardware analysis or functional analysis. Hardware analysis is a bottom-up approach, and functional analysis is a top-down approach.

Guidelines are given in these two documents as to how the approaches might be more effectively applied to a system design to obtain the maximum benefit from the procedures. This involves simpler procedures (FMECA) early in the design development, and more refined analysis (fault tree) as the design progresses.

Military Aircraft. Probabilistic risk assessment for military aircraft is conducted according to U.S. Department of Defense Mil-Std-882B (1984). This document outlines the process for identifying the systems that have unacceptable levels of risk. The process has three steps, which are illustrated in the following three figures, and they are based on Section 4.5 of the Military Specification. Each element in the system must first be assigned a hazard probability level, a number between 1 and 6. It represents the frequency of failure for the item under consideration as shown in Fig. A-2. Second, a hazard level is assigned. The hazard level is between 1 and 4 and represents the severity of the failure as shown in Fig. A-3. Third, the product of the previous values is found. It will be between 1 and 24 and is known as the real hazard index. Figure A-4 shows how this is used to identify areas of unacceptable risk. Any item with a real hazard index greater than 12 must be modified in some way to reduce the risk level.

A3.2 □ PROBABILISTIC FRACTURE MECHANICS ANALYSIS

Probabilistic fracture mechanics methods are used in some areas, but there is no formal requirement. The U.S. Air Force is beginning to consider the potential of probabilistic design methods, but no formal requirements have yet been made.

	Level	Specific Individual Item	Fleet (or Inventory)	Approximate Number of Occurrences per Flight Hour
Frequent	6	Likely to Occur Frequently	Continuously Experienced	More Than 10 per 10,000 Flight Hours
Reasonably Probable	5	Will Occur Several times in Life of Individual Item	Will Occur Frequently	Between 1 and 10 per 10,000 Flight Hours
Occasional	4	Unlikely to Occur in Life of Specific Item	Will Occur Several Times	Between 1 and 10 per 100,000 Flight Hours
Remote	3	So Improbable That it Can Be Assumed That it Will Not Happen	Unlikely to Occur But Possible	Between 1 and 10 per 1,000,000 Flight Hours
Extremely Improbable	2	Probability of Occurrence Cannot Be Distinguished From Zero	So Improbable That it Can Be Assumed That the Fleet Will Never Experience it	Less Than 1 1,000,000 Flight Hours
Impossible	1	Physically Impossible to Occur		0

GP03-0007-460-D/kt

FIG. A2 HAZARD PROBABILITY LEVELS
[Adapted from U.S. Department of Defense Mil-Std-882B (1984)]

Category	Hazard Level	Description	
I	4	Catastrophic	May Cause Death or System Loss
II	3	Critical	May Cause Severe Injury, Severe Occupational Illness, or Major System Damage
III	2	Marginal	May Cause Minor Injury, Minor Occupational Illness, or Minor System Damage
IV	1	Negligible	Will not Result in Injury, Occupational Illness, or System Damage

GP03-0007-461-D/kt

FIG. A3 HAZARD LEVEL CATEGORY
[Adapted from U.S. Department of Defense Mil-Std-882B (1984)]

Hazard Level, HL	Probability Level, PL 6	5	4	3	2	1
4	24	20	16	12	8	4
3	18	15	12	9	6	3
2	12	10	8	6	4	2
1	6	5	4	3	2	1

Real Hazard Index (RHI) = HL x PL

RHI	Criteria
1-6	Acceptable Without Review
8-10	Acceptable With Review
12-24	Unacceptable

FIG. A4 REAL HAZARD INDEX
[Adapted from U.S. Department of Defense Mil-Std-882B (1984)]

A3.3 □ INSPECTIONS

What to Inspect. The structural items that are inspected are determined from the stress analysis of the airframe. This decision includes results from finite element analysis, the nature and location of structural details (holes, fillets, etc.), and experience gained in previous aircraft programs. A set of "critical points" are determined for the aircraft. They become the focus of attention during inspections.

When to Inspect. The frequency of inspections is dictated by the Military Specification and depends on the portion of life used. Life depends not only on the hours of flight time, but also on the severity of the use of the aircraft. Crack growth projection is used to determine the life of the aircraft.

One of the main causes of failure in an aircraft structure is the growth of cracks from holes. There are many fastener holes on an aircraft (an Air Force F-15 Air Superiority Fighter, for example, has over 300,000). Frequently, the areas that need to be inspected are inside the wing, tail, or fuselage structure. Inspection necessitates the removal of the outer skin. The removal and reinstallation of the skin can cause damage to the fastener holes. This procedure can precipitate failures. A delicate balance exists, consequently, between too many and too few inspections.

A3.4 □ NEW DEVELOPMENTS

Tactical fighter aircraft are designed for a finite lifetime using a baseline or design, severity of use spectrum. This spectrum represents the typical mission profile for the aircraft. Aircraft that are flown very hard (many high "g" maneuvers) expend their design life much faster than aircraft that see a benign flight spectrum (few high "g" maneuvers). The rate of increase or decrease in life expenditure from that predicted for the design spectrum is directly proportional to the severity of the usage and is referred to as the usage factor. The flight hours of each aircraft must be multiplied by its usage factor prior to making any comparisons to historical data from other aircraft.

When making predictions of risk of structural failure, the usage factor must be accounted for prior to determining the statistical distributions. Three parameter Weibull distributions are used to model both the failure distribution and the aircraft flight hour distribution. The failure distribution must be determined from fleet inspection data and is generated for each critical (or inspection) point in the aircraft. The probability of failure is determined from the two Weibull distributions, as shown in Fig. A-5, which is based on work by Christian, et al. (1986), Saff, et al. (1987), and Smith, et al. (1990). The shaded area indicates a finite probability of failure whose magnitude is a function of the degree of overlap of the two distributions.

Aircraft are inspected in the field at a number of critical points in the structure. The critical points are presently identified by the aircraft manufacturer based on the assessment of load data from finite element models; consideration of structural details, such as holes and cutouts; and past experience. Detailed analysis and testing are performed for each of these critical locations to produce a crack growth curve. The crack growth curve is used to help build the aircraft failure distribution for the specific critical point of interest.

A field inspection of a critical point yields either no crack, a crack of noncritical size, or a crack of critical size. "No crack" data points enter the analysis as suspended items, critical size cracks are treated as failures, and noncritical cracks are analytically projected to critical size using the crack growth curve for that location (Fig. A-6), which is also based on work by Christian, et al. (1986), Saff, et al. (1987), and Smith, et al. (1990). These analytically determined failure lives are used along with the lives of any existing critical flaws to determine the failure distribution. The distribution of aircraft lives is found from the hours on each aircraft modified by its unique usage factor.

With the two distributions determined, the fleet commander can perform a number of different analyses to assist in the management of the aircraft fleet. Maintenance and inspections can be scheduled, the significance of inspection data can be assessed, and the risk of failure for additional flight hours and modified usage can be determined.

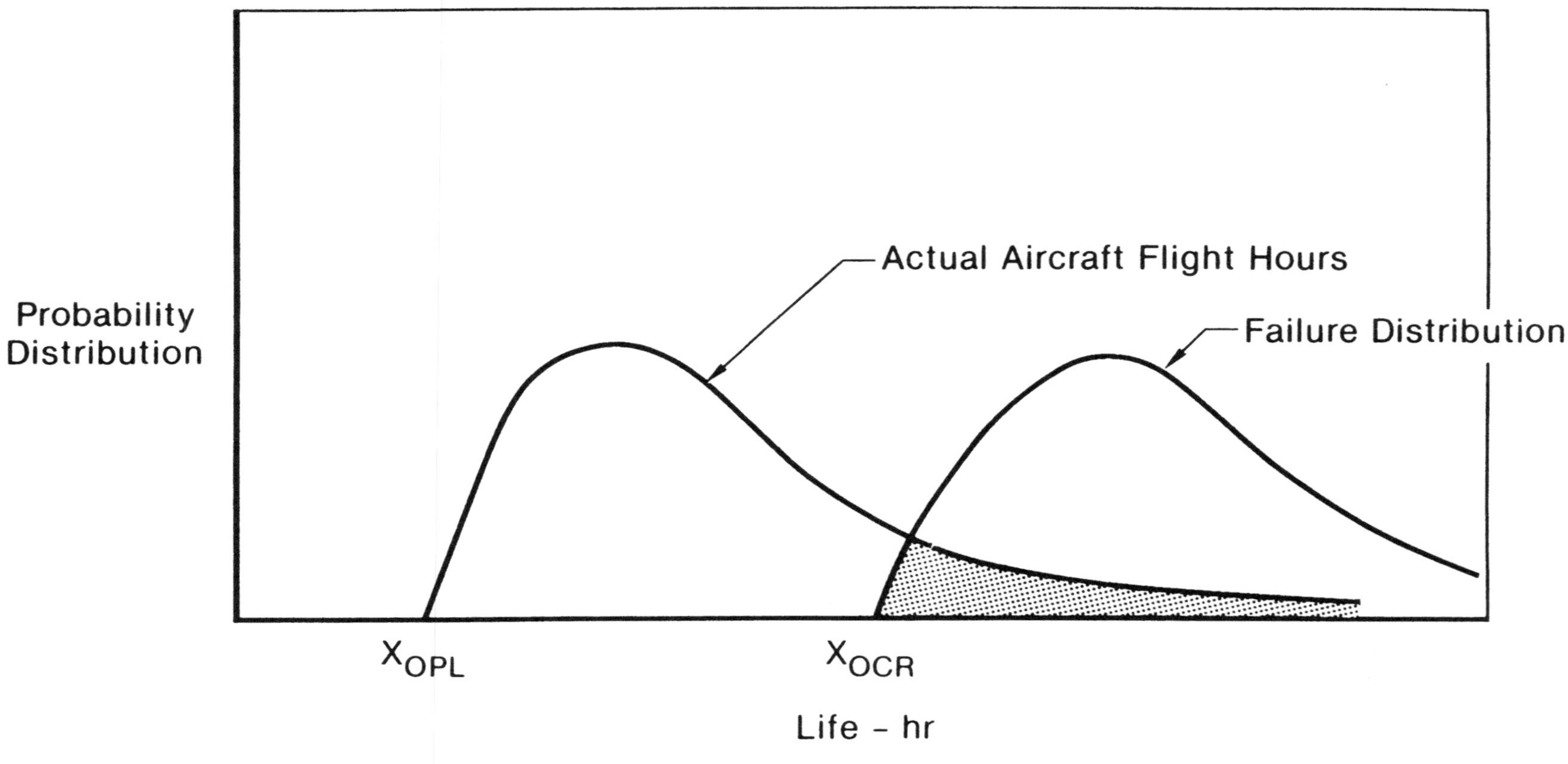

FIG. A5 PROBABILITY OF FAILURE FROM TWO WEIBULL DISTRIBUTIONS
[Based on Christian, et al. (1986), Saff, et al. (1987), and Smith, et al. (1990)]

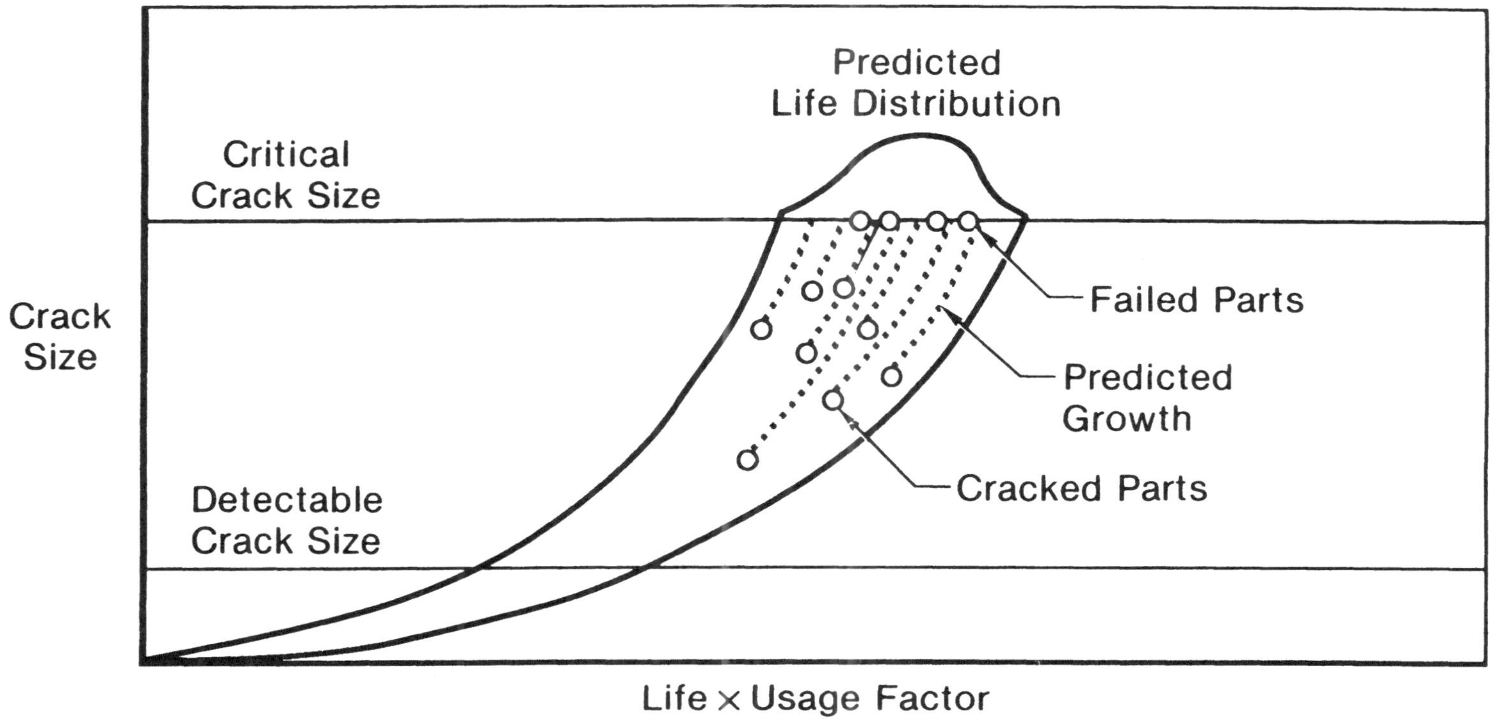

FIG. A6 PREDICTED LIFE DISTRIBUTION USING CRACK GROWTH CURVE
[Based on Christian, et al. (1986), Saff, et al. (1987), and Smith, et al. (1990)]

CIVIL ENGINEERING AND MARINE STRUCTURES

A4

This section provides a summary of selected studies on reliability-based inspection and risk-benefit analysis mainly in the civil and marine industries.

Mohammadi and Longinow (1985) described the importance of, and the necessity for, a probability-based design practice for highway bridges. They emphasized the significance of a probability analysis in determining the frequency of inspection of highway bridges. A study by Raczon (1987) showed that states had no written procedures for varying the scope or frequency of inspections. The 1987 Surface Transportation and Uniform Relocation Assistance Act required that the Federal Highway Administration continue to specify bridge inspection methods, the maximum time period between inspections and qualifications of inspectors. Also, a national certification program for bridge inspectors was suggested.

The National Transportation Safety Board (NTSB) (1984) determined that the probable cause of the 1983 collapse of a suspended span of Interstate Route 95 highway bridge over the Mianus River in Connecticut was due to the lateral displacement of the hangers of the pin-and-hanger suspension assembly of the span by corrosion-induced forces. The condition was not detected due to deficiencies in the State of Connecticut's bridge safety inspection and maintenance program.

In 1986, the board determined that the probable cause of the collapse of the US 43 Chickasawbogue Bridge spans in Alabama in 1985 was not detected due to the inadequate inspection of the underwater bridge elements by the State of Alabama, as given in an NTSB investigation (1985). Also, in 1988 the causes of the 1987 collapse of a 5-span, highway bridge in New York were investigated by the NTSB (1988). The report discussed deficiencies uncovered in the bridge inspection programs of the New York State Thruway Authority and the New York State Department of Transportation. Recommendations were proposed to revise existing guidelines for design, maintenance, and inspection of bridges. In addition, it was recommended that the U.S. Department of Transportation Inspector General should periodically review the Federal Highway Administration bridge inspection audit program for compliance with the National Bridge Inspection Standards.

Imbsen and Schamber (1984) proposed a methodology for rating reinforced-concrete bridges. The methodology was presented in a reliability-based format by using approximate load and resistance factors. By using this format, the probability theory and engineering condition can be rationally combined to allow for independent consideration of each of the major variables that can affect the determination of the load capacity of a bridge. This methodology includes consideration of the level of effort in maintenance and inspection, degree of load-limit enforcement, and refinement used in simulating the bridge.

Knepp, et al. (1976) performed a study on the benefit-risk analysis in the design of highway stream crossing. In this model, the total costs associated with alternative designs were estimated, weighting factors were established for the costs according to the probability of occurrence of adverse events, and decision criteria were established. Harrison and Grenke (1983) and Cottrell (1988) provided methods for the selection of the frequencies of inspection of highway safety hardware based on their accident history and the level of service to be provided. It was concluded that the procedure is a useful method for determining highway safety hardware maintenance guidelines.

Crist (1986) presented a strategy to develop a preliminary inspection frequency model and the requirements needed to determine the order of inspection for waterfront facilities. The criteria used for determining when the inspections should be performed were construction material, facility age, present condition, facility environment, and mission requirements. Fujimoto, et a1. (1989) developed a method based on Bayesian reliability analysis that estimates the optimal inspection schedule of structures with multiple components using the data collected during inservice inspections. Mohammadi and Yazbeck (1989) developed a method based on probabilistic modeling for estimating the critical time intervals between inspections for highway bridges.

In the area of probabilistic fracture mechanics, models were used to determine the risk of failure of highway steel bridges, and establish inspection and maintenance strategies [Yazdani, et al. (1987) and Tallin (1988)]. A similar model was used for railroad rails (Kesling and Whittaker, 1985). A probabilistic model for fatigue crack growth had been applied by Madsen (1987) for offshore structures that accounts for uncertainties in loading, initial defects, critical crack size, and material parameters, and in the computation of the stress intensity factor.

Shirole and Hill (1978) suggested a systems approach to the bridge structure replacement-priority planning. Structural condition, functional adequacy, safety, essentiality to traffic, and other criteria for setting replacement priority were developed and evaluated. Budgetary, environmental, development policy, and other constraints on the replacement priority were identified and analyzed for their possible impact. A quantitative methodology was developed, based on assignment of weights to the rated criteria. Shanafelt and Horm (1984) provided a methodology for a decision-making process in the evaluation and repair of damaged steel bridge members. The methodology assembles information concerning the effect of these repair techniques on the service life, safety, performance, and maintenance of the structure. Decisions on method of repair must also consider the cost, user inconvenience, and aesthetics of the repair technique. Other researchers used risk analysis as a tool in selection between alternatives [Peterson (1986)].

Inspection and maintenance methods have large effects on structural life expectancy and extension. Many other factors affect the life expectancy of a structure. Kitagawa (1985) discussed selected life prediction methods. They include life prediction by the fracture process models, life prediction based on degradation of materials, and optimization of inspection periods based on life prediction using single crack growth models. In bridge structures, Parekh, et al. (1986), and Berger and Gordon (1978) suggested methods for life extension of bridge structures by providing a posting policy. The policy was based on minimizing risk and maximizing benefits to the users. Moses and Verma (1987) suggested guidelines for the evaluation of existing bridges. It was primarily intended to replace the present provisions in the American Association of State Highway and Transportation Officials (AASHTO) Manual for Maintenance and Inspection of Bridges pertaining to the bridge posting calculations. A methodology of structural life assessment was suggested by Ayyub, White, and Purcell (1989) for marine structures. The methodology was based on probabilistic analysis using reliability concepts [Ayyub and Haldar (1984) and White and Ayyub (1985 and 1987)] and the statistics of extremes. The methodology resulted in the probability of failure of the structural system according to the identified failure modes as a function of time, i.e., structural life. The results can be interpreted as the cumulative probability distribution function (CDF) of structural life. Due to the unknown level of statistical correlation between failure modes, limits or bounds on the CDF of the structural life were established. The limits correspond to the extreme cases of fully correlated and independent failure modes.

For example, the CDFs of structural life of an example marine vessel were determined for two inspection strategies; namely, inspection every year and inspection every two years with a warranty inspection at the end of the first year. Example results are shown in Fig. A-7.

In the area of civil engineering, probabilistic fracture mechanics models were used to determine the risk of failure of steel highway bridges and to establish inspection and maintenance strategies. Other examples include reliability-based design and inspection of debris basins (for mudflows), public transportation systems, pipeline networks, etc.

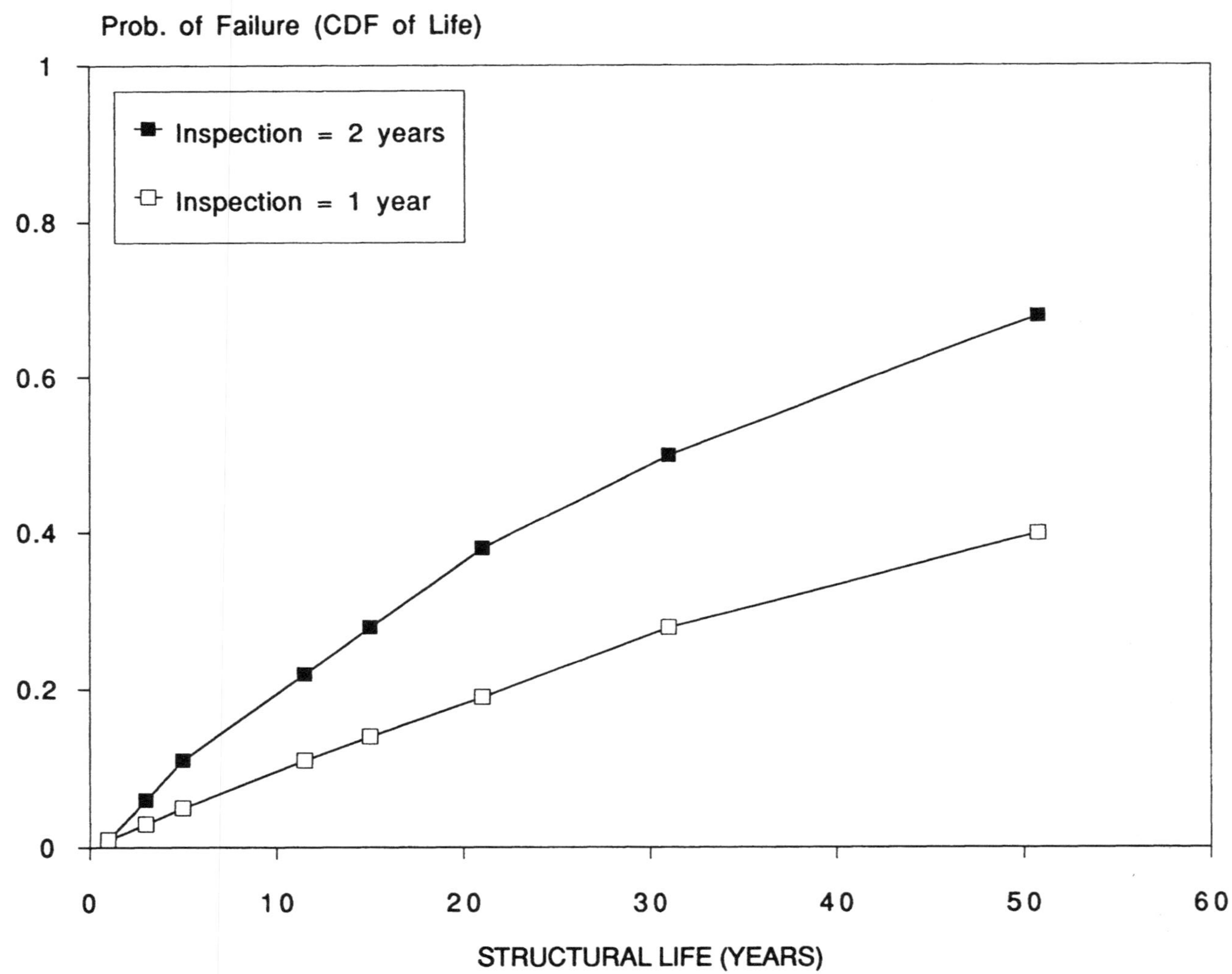

FIG. A7 LOWER LIMIT ON STRUCTURAL LIFE BASED ON PLATE DEFORMATION
[Based on Ayyub and Haldar (1984), and White and Ayyub (1985) and (1987)]

INDUSTRIAL INSURANCE A5

The objective of inspections of insured locations is to perform effective, efficient, and economical risk assessments, and consequential loss prevention and property preservation through sound engineering practices. This pertains to the identification of property loss exposures attributed to the existence of overt or covert hazards related to the processes and facilities under examination, or hazards associated with planned or implemented changes.

Usually the inspection covers the entire plant, with particular attention given to conditions dictated by the following considerations:

- insurance coverage
- applicable deductibles
- exposures having the largest insured loss potential otherwise defined as critical and important exposures (see definitions below)
- jurisdictional requirements

The inspection frequency is determined by the following factors:

(a) the type of inspection (i.e., planned regular inspections, or one time special inspections);

(b) jurisdictional requirements;

(c) exposure group classification; based on the probable maximum loss (net of property damage and business interruption deductible amounts), occupancies are classified in groups requiring semiannual, annual, or one time only inspection (unless other jurisdictional regulations apply).

Probable maximum loss (PML) is defined as the largest loss that is likely to occur after a structural failure (on the basis of past experience and design of the equipment). Maximum foreseeable loss (MFL) is the largest loss that may be expected from a structural failure on the basis of past experience and design of the equipment. Critical exposures are those that present a PML or MFL larger than an amount specified by applicable insurance policies. Tabulated values of PML and MFL based on past experience provide assistance and guidance for the identification of critical and important exposures.

During inspections, deviations from acceptable engineering practices and loss prevention principles are identified and appropriate recommendations are issued. The basis for objective judgment for such recommendations includes Loss Prevention Standards (reflecting sound engineering analysis), analysis of loss experience, and, lately, probabilistic modeling. In many cases loss prevention recommendations are supported by specialized testing to determine particular failure modes, special hazards, and the needed protection. Such testing includes the following:

- combustion tests
- explosion tests
- sensitivity tests (smoke detection systems, low water level systems, etc.)
- metallurgical tests (metallurgical examination, chemical analysis, mechanical testing, nondestructive examination)

- reliability, malfunction, and failure examinations

Probabilistic engineering risk-analysis methods have been employed in progressive branches of industrial insurance to assess efficiently and economically the risks associated with loss prevention and property conservation. Karydas (1987) shows an application of such methods to the occurrence rate of boiler failures caused by undetected low water level conditions in watertube power boilers, as discussed below.

Conventional control and safety systems have been compared with programmable electronic systems. The reduction in the expected number of undetected low water level conditions with increasing frequency of inspection and preventive maintenance was quantified (see Figs. A-8 and A-9). The applied methodology included failure modes, effects, and criticality analysis; fault tree analysis; and uncertainty analysis. Sources for component failure rates included insurance loss data and expert opinion. The results of the study indicated, among other things, that inspection frequency of the control system considerably affects the expected number of failures (ENF) within specified time periods. Quarterly versus annual maintenance provides more than a factor of six reduction of failure probability for a low water fuel cutoff (LWFC) system that includes a sightglass water level monitoring system. Further improvements in ENF are obtained by installing a second LWFC system or by substituting a conventional LWFC system (with a rate of 769 failures per million hours) with electronic systems demonstrating proven lower failure rates.

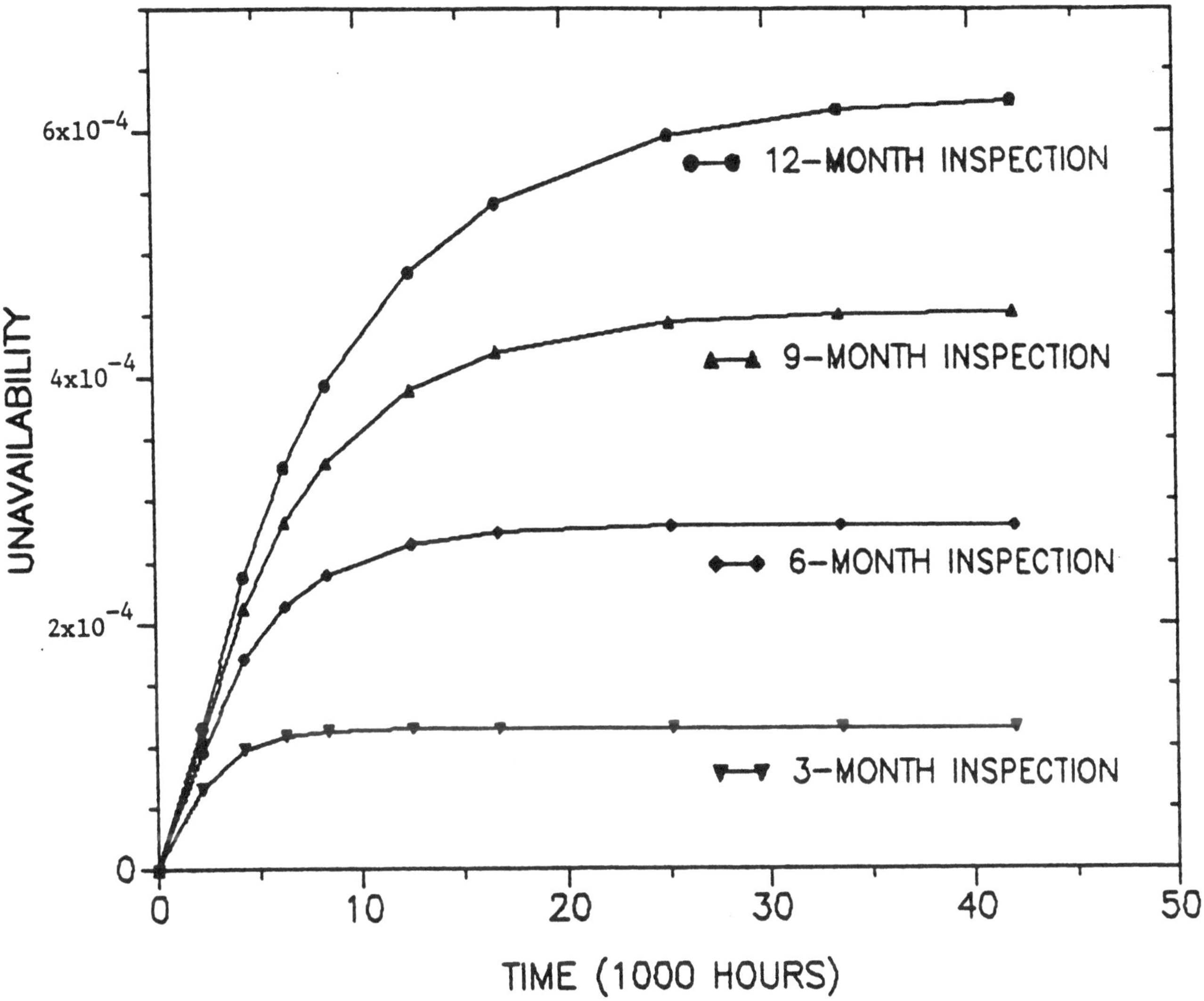

FIG. A8 WATER TUBE BOILER LOW WATER LEVEL PROBABILITY AS A FUNCTION OF INSPECTION INTERVAL
[Karydas (1987)]

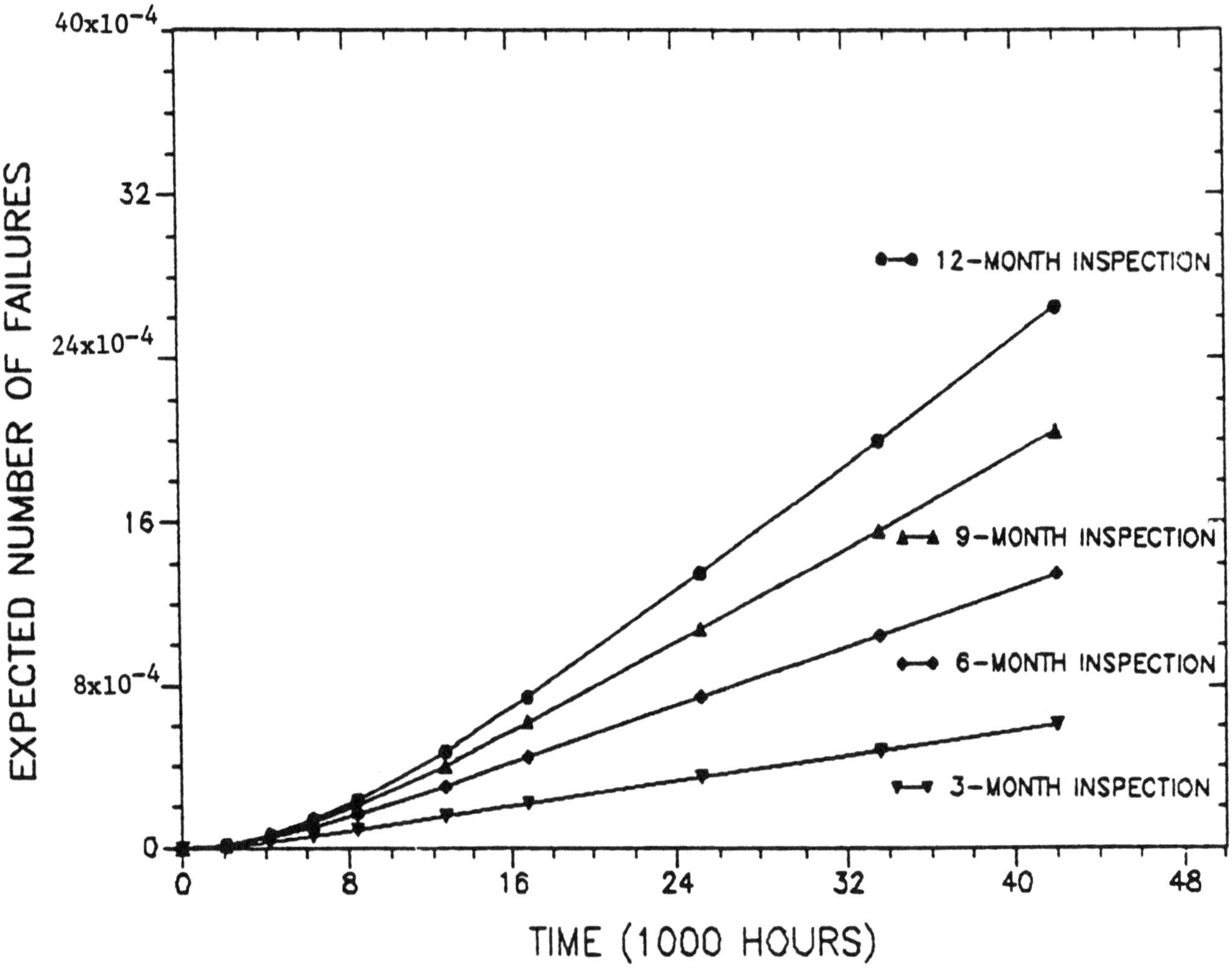

FIG. A9 WATER TUBE BOILER EXPECTED NUMBER OF FAILURES AS A FUNCTION OF INSPECTION INTERVALS
[Karydas (1987)]

APPENDIX B
MATHEMATICAL EXPRESSIONS FOR TABLES 2-7 THROUGH 2-10

TABLE B1
POINT ESTIMATE METHOD FOR SINGLE FAILURE MODE
(See Table 2-7)

	A	B	C
1	System	Element	Probability
2			of Faliure
3	1		
4		1	0.00000000000001
5		2	0.00000000000001
6		3	0.00000000000001
7		4	0.00001
8		5	0.00001
9		6	0.00001
10		7	0.01
11		8	0.01
12		9	0.01
13	Upper Limit		=1-((1-C4)*(1-C5)*(1-C6)*(1-C7)*(1-C8)*(1-C9)*(1-C10)*(1-C11)*(1-C12))
14	Lower Limit		=MAX(C4:C12)
15	Average		=SQRT(C13*C14)
16	2		
17		1	0.0000000001
18		2	0.0000001
19		3	0.00000001
20		4	0.00000001
21		5	0.0000001
22		6	0.001
23		7	0.000001
24		8	0.00000001
25		9	0.000000001
26	Upper Limit		=1-((1-C17)*(1-C18)*(1-C19)*(1-C20)*(1-C21)*(1-C22)*(1-C23)*(1-C24)*(1-C25))
27	Lower Limit		=MAX(C17:C25)
28	Average		=SQRT(C26*C27)

TABLE B1 (CONT'D)
POINT ESTIMATE METHOD FOR SINGLE FAILURE MODE
(See Table 2-7)

	D	E	F	G	H	I
1		Magnitude of			Ranking according to	
2	Casualties	Damage	Human Risk	Economic Risk	Human Risk	Economic Risk
3						
4	0	0	=C4*D4	=C4*E4	8	8
5	1000	1000	=C5*D5	=C5*E5	7	7
6	1000000	1000000	=C6*D6	=C6*E6	6	6
7	1	1	=C7*D7	=C7*E7	5	5
8	1000	1000	=C8*D8	=C8*E8	4	4
9	1000000	1000000	=C9*D9	=C9*E9	2	2
10	0	0	=C10*D10	=C10*E10	8	8
11	1000	1000	=C11*D11	=C11*E11	2	2
12	1000000	1000000	=C12*D12	=C12*E12	1	1
13						
14						
15			=SUM(F4:F12)	=SUM(G4:G12)	1	1
16						
17	0	1000	=C17*D17	=C17*E17	8	7
18	1000	100	=C18*D18	=C18*E18	4	4
19	1000000	5000	=C19*D19	=C19*E19	2	3
20	1	1000	=C20*D20	=C20*E20	7	4
21	1000	1	=C21*D21	=C21*E21	4	7
22	1000000	0	=C22*D22	=C22*E22	1	9
23	0	1000000	=C23*D23	=C23*E23	8	1
24	1000	1000000	=C24*D24	=C24*E24	6	2
25	1000000	1000	=C25*D25	=C25*E25	3	6
26						
27						
28			=SUM(F17:F25)	=SUM(G17:G25)	2	2

TABLE B2
INTERVAL ANALYSIS FOR SINGLE FAILURE MODES
(See Table 2-8)

	A	B
1	System	Element
2		
3	1	
4		1
5		2
6		3
7		4
8		5
9		6
10		7
11		8
12		9
13	Upper Limit	
14	Lower Limit	
15	Best Estimate	
16	2	
17		1
18		2
19		3
20		4
21		5
22		6
23		7
24		8
25		9
26	Upper Limit	
27	Lower Limit	
28	Best Estimate	

TABLE B2 (CONT'D)
INTERVAL ANALYSIS FOR SINGLE FAILURE MODES
(See Table 2-8)

	C	D
1	Probability	of
2	Lower	Best Estimate
3		
4	0.000000000000000001	0.000000000000001
5	0.0000000000000001	0.000000000000001
6	0.0000000000000001	0.000000000000001
7	0.000000001	0.00001
8	0.000000000001	0.00001
9	0.000001	0.00001
10	0.000000001	0.01
11	0.00001	0.01
12	0.001	0.01
13	=1-((1-C4)*(1-C5)*(1-C6)*(1-C7)*(1-C8)*(1-C9)*(1-C10)*(1-C11)*(1-C12))	
14	=C12	
15	=C14	
16		
17	0.0000000000001	0.0000000001
18	0.0000000001	0.0000001
19	0.0000000001	0.00000001
20	0.0000000001	0.00000001
21	0.00000001	0.0000001
22	0.0001	0.001
23	0.0000001	0.000001
24	0.0000000000001	0.00000001
25	0.00000000000001	0.000000001
26	=1-((1-C17)*(1-C18)*(1-C19)*(1-C20)*(1-C21)*(1-C22)*(1-C23)*(1-C24)*(1-C25))	
27	=C22	
28	=C27	

TABLE B2 (CONT'D)
INTERVAL ANALYSIS FOR SINGLE FAILURE MODES
(See Table 2-8)

	E	F	G
1	Failure	Magnitude	of
2	Upper	Lower	Best Estimate
3			
4	0.0000000001	0	0
5	0.00000001	100	1000
6	0.000001	1000	1000000
7	0.001	0	1
8	0.0001	500	1000
9	0.0001	1450	1000000
10	0.1	0	0
11	0.01	10	1000
12	0.1	500	1000000
13	=1-((1-E4)*(1-E5)*(1-E6)*(1-E7)*(1-E8)*(1-E9)*(1-E10)*(1-E11)*(1-E12))		
14	=E12		
15	=E13		
16			
17	0.0000001	0	0
18	0.0000001	2	1000
19	0.0001	200	1000000
20	0.00000011	0	1
21	0.0001	900	1000
22	0.01	100000	1000000
23	0.0001	0	0
24	0.00001	500	1000
25	0.00001	100000	1000000
26	=1-((1-E17)*(1-E18)*(1-E19)*(1-E20)*(1-E21)*(1-E22)*(1-E23)*(1-E24)*(1-E25))		
27	=E22		
28	=E26		

TABLE B2 (CONT'D)
INTERVAL ANALYSIS FOR SINGLE FAILURE MODES
(See Table 2-8)

	H	I	J	K	L	M
1	Casualties	Magnitude	of	Damage	Human Risk	
2	Upper	Lower	Best Estimate	Upper	Lower	Upper
3						
4	0	0	0	0	=C4*F4	=E4*H4
5	100000	200	1000	2000	=C5*F5	=E5*H5
6	2000000	5000	1000000	1500000	=C6*F6	=E6*H6
7	100	0	1	100	=C7*F7	=E7*H7
8	2000	500	1000	5000	=C8*F8	=E8*H8
9	1500000	99999	1000000	1900000	=C9*F9	=E9*H9
10	0	0	0	0	=C10*F10	=E10*H10
11	2000	300	1000	7000	=C11*F11	=E11*H11
12	1100000	10000	1000000	1800000	=C12*F12	=E12*H12
13						
14						
15					=SUM(L4:L12)	=SUM(M4:M12)
16						
17	0	1	1000	1200	=C17*F17	=E17*H17
18	3000	50	100	1000	=C18*F18	=E18*H18
19	1100000	1000	5000	6000	=C19*F19	=E19*H19
20	100	20	1000	5000	=C20*F20	=E20*H20
21	1200	0	1	100	=C21*F21	=E21*H21
22	1100000	0	0	0	=C22*F22	=E22*H22
23	0	100000	1000000	2000000	=C23*F23	=E23*H23
24	7000	1000	1000000	2000000	=C24*F24	=E24*H24
25	1100000	999	1000	100000	=C25*F25	=E25*H25
26						
27						
28					=SUM(L17:L25)	=SUM(M17:M25)

TABLE B2 (CONT'D)
INTERVAL ANALYSIS FOR SINGLE FAILURE MODES
(See Table 2-8)

	N	O	P	Q
1	Damage Risk		Ranking according to	
2	Lower	Upper	Human Risk	Economic Risk
3				
4	=C4*I4	=E4*K4	8	8
5	=C5*I5	=E5*K5	7	7
6	=C6*I6	=E6*K6	4	4
7	=C7*I7	=E7*K7	6	6
8	=C8*I8	=E8*K8	5	5
9	=C9*I9	=E9*K9	2	2
10	=C10*I10	=E10*K10	8	8
11	=C11*I11	=E11*K11	3	3
12	=C12*I12	=E12*K12	1	1
13				
14				
15	=SUM(N4:N12)	=SUM(O4:O12)	1	1
16				
17	=C17*I17	=E17*K17	8	7
18	=C18*I18	=E18*K18	6	8
19	=C19*I19	=E19*K19	2	4
20	=C20*I20	=E20*K20	7	6
21	=C21*I21	=E21*K21	4	5
22	=C22*I22	=E22*K22	1	9
23	=C23*I23	=E23*K23	8	1
24	=C24*I24	=E24*K24	5	2
25	=C25*I25	=E25*K25	3	3
26				
27				
28	=SUM(N17:N25)	=SUM(O17:O25)	2	2

TABLE B3
POINT ESTIMATE METHOD FOR MULTIPLE FAILURE MODES
(See Table 2-9)

	A	B	C	D	E	F	G
1	Element	Failure	Probability	Magnitude of			
2		Mode	of Faliure	Casualties	Damage	Human Risk	Economic Risk
3	1	Mode 1	0.0000000000001	0	200	=C3*D3	=C3*E3
4		Mode 2	0.0001	0	100	=C4*D4	=C4*E4
5	2	Mode 1	0.0000001	100000	10000	=C5*D5	=C5*E5
6	3	Mode 1	0.0000000001	1000	3500	=C6*D6	=C6*E6
7		Mode 2	0.00000001	300	200	=C7*D7	=C7*E7
8	4	Mode 1	0.00000001	34	700	=C8*D8	=C8*E8
9		Mode 2	0.000000000001	0	500	=C9*D9	=C9*E9
10		Mode 3	0.00000001	50	90	=C10*D10	=C10*E10
11	5	Mode 1	0.00001	750	45	=C11*D11	=C11*E11
12	6	Mode 1	0.0000000001	4000	430	=C12*D12	=C12*E12

	H	I
1	Ranking according tc	
2	Human Risk	Economic Risk
3	8	10
4	8	1
5	1	2
6	7	7
7	3	5
8	6	4
9	8	9
10	4	6
11	2	3
12	5	8

TABLE B4
INTERVAL ANALYSIS FOR MULTIPLE FAILURE MODES
(See Table 2-10)

	A	B	C	D	E	F	G	H
1	Element	Failure Mode						
2						Casualties		
3			Min	Best Estimate	Max	Min	Best Estimate	Max
4	1	Mode 1	0.0000000000000001	0.0000000000001	0.00000001	0	0	30
5		Mode 2	0.000001	0.0001	0.001	0	0	0
6	2	Mode 1	0.000000001	0.0000001	0.00001	3000	100000	500000
7	3	Mode 1	0.000000000001	0.0000000001	0.0000001	30	1000	50000
8		Mode 2	0.0000000001	0.00000001	0.000001	200	300	350
9	4	Mode 1	0.000000001	0.00000001	0.00001	20	34	500
10		Mode 2	0.000000000001	0.000000000001	0.000000001	0	0	12
11		Mode 3	0.000000000001	0.00000001	0.000001	40	50	60
12	5	Mode 1	0.0000000001	0.00001	0.0001	500	750	70000
13	6	Mode 1	0.0000000000001	0.0000000001	0.000000001	2000	4000	80000

	I	J	K	L	M	N	O	P	Q
1	Magnitude o							Ranking according to	
2	Damage			Human Risk		Economic Risk		Human Risk	Economic Risk
3	Min	Best Estimate	Max	Min	Max	Min	Max		
4	70	200	3000	=C4*F4	=E4*H4	=C4*I4	=E4*K4	8	9
5	90	100	150	=C5*F5	=E5*H5	=C5*I5	=E5*K5	10	2
6	2000	10000	100000	=C6*F6	=E6*H6	=C6*I6	=E6*K6	2	1
7	700	3500	4000	=C7*F7	=E7*H7	=C7*I7	=E7*K7	4	7
8	20	200	500	=C8*F8	=E8*H8	=C8*I8	=E8*K8	5	6
9	500	700	7000	=C9*F9	=E9*H9	=C9*I9	=E9*K9	3	3
10	400	500	550	=C10*F10	=E10*H10	=C10*I10	=E10*K10	9	10
11	10	90	100	=C11*F11	=E11*H11	=C11*I11	=E11*K11	7	8
12	30	45	300	=C12*F12	=E12*H12	=C12*I12	=E12*K12	1	4
13	400	430	1000000	=C13*F13	=E13*H13	=C13*I13	=E13*K13	6	5

APPENDIX C
STATISTICAL APPROACH TO THE ANALYSIS OF ISI DATA USING THE BAYES METHOD
[O. J. V. Chapman (1983)]

A Statistical Approach to the Analysis of ISI Data Using the Bayes Method

O.J.V. Chapman

Rolls-Royce and Associates Limited, P.O. Box 31, Derby DE2 8BJ, U.K.

ABSTRACT

This paper describes a means of determining the effectiveness of In-Service Inspection (ISI). The objective being to optimise any ISI programme and thus minimise downtime and man dose exposure in nuclear application.

The report proposes a statistical method for handling information on operating life and ISI results accrued to date and from planned ISI programmes in future. The method makes the assumption that a family group of components is amenable to a statistical representation. The family group may extend across several different reactors and stations provided the area being inspected is common, has been built to the same standard and operated in a similar manner. For each area the input required for the prior assessment is a family of initial crack size distributions, crack densities and crack growth functions. These provide an initial probability of failure distribution.

A Bayes approach is used to update the prior probability of failure distribution using the results of the ISI programme. As the ISI data builds up, the probability of failure distribution for that particular component or ISI area approaches the true situation. This interpretation of the ISI results will allow better optimisation of future ISI programmes.

At present the method is being applied to a group of reactor pressure vessels which are of a similar design and have been built to a common standard. It is hoped to both gain confidence in the integrity of the vessels as a group and optimise the ISI programme between various detailed parts of the vessel.

1. Introduction

This paper describes a probabilistic model to define the benefits of ISI.

The model quantifies the value of ISI in terms of the increasing confidence successive inspections provide in showing that the required or 'target' probability of failure assigned to a vessel or area of a vessel is or is not being realised. A range of different defect densities and distributions are used as initial input to describe possible start of life conditions for a vessel area and a range of different crack growth rates are defined to fully embrace possible defect behaviour in service. The results of successive ISI's are interpreted and the reality of the situation as to the failure probability appropriate to that area is indicated by a converging solution.

The model recognises the benefit of plant operating years accrued to date and the value of in service inspections performed to date. The model can be developed to show the relative importance of one vessel or component to another within the reactor plant as well as the relative importance of different areas within a given vessel.

A significant attribute of the model is its ability to provide a 'running statement' on the benefit of ISI as each inspection is performed and the relative value of inspecting a greater or lesser proportion of a given area. It can thus indicate subsequent ISI needs and enable future programmes to be optimised by balancing the inspection requirements such that all areas accumulate similar levels of confidence against their respective failure probability targets.

2. Basis of New Method

The underyling basis of this approach is fundamentally quite simple. It lies in asking the following question: "Given that we believe a certain situation exists, what would we expect to find during an ISI?". Given then the outcome of an ISI, the follow up question becomes: "How does this knowledge affect the original belief, or confidence in that situation?". Note that the question is not, how does ISI improve the situation but how does it affect ones confidence in a given situation. Whilst it must be true that ISI will tend to reduce the probability of failure from some unknown level, the modelling in this paper does not set out to demonstrate this or to take any advantage of it. The over-riding objective is to use ISI to demonstrate that the situation is one that is acceptable.

3. Probability of Failure

We must now define what is meant by acceptable. Clearly in the context of a nuclear plant this must be related to the probability of leakage. (Catastrophic failure is considered as a subset of leakage). Given that there is some level at which this probability is acceptable, then the objective is to demonstrate that the vessel or component being inspected is at least as good, if not better, than this level.

The ingredients required to determine the probability of leakage are :-

1)	Defect denstiy	'D'
2)	Defect size distribution	f(a)
3)	Crack growth	g(a)
4)	Failure distribution	h(a)

To use this data in a deterministic analysis would require it to be known within reasonably close tolerances. This can only be done with a considerable effort, and there are many who would say it can never be done. However even those that say it cannot be done must concede that in principle, at least, the statement is correct and that it is even possible to establish the form of these parameters. It is the assigning of absolute values that is often the bone of contention.

4. Estimating the Parameters

Whilst the above parameters will not be known with any great confidence, it can be assumed that there is a reasonable knowledge of the general form and scale of these parameters. For example the defect size distribution is generally accepted to be some form of decaying exponential. An alternative is the Weibull function, which can be made the same as the exponential function but can also give better control over the tail of f(a).

The same applies for the crack growth function, g(a). There is a wealth of knowledge on crack growth itself and a reasonable knowledge on the effect of the reducing ligament as the crack depth approaches a through thickness defect. Therefore a reasonable knowledge of the type and extent of the loading conditions permits a good estimate of at least the form of g(a), if not its actual value. With regard to the defect density 'D', this is always a problem but once again experience enables some respectable guesses to be made.

This leaves h(a) to be determined. In theory it would be possible to estimate a distribution of h(a). However it is more practical to think of a defect size above which we can consider failure to have occurred. This value could be set at the thickness but the possibility of snap through, stable or unstable, together with the possible increase in crack growth as the ligament size decreases to zero, leads one to choose a value for h(a) that is less than the thickness. Therefore a single value for h(a) can be chosen such that $h(a) = a_c <$ thickness.

The preceeding suggests that we can estimate the probability of failure for a given situation. But the argument could be turned around, that is, given that there is an acceptable probability of failure, Pf_i what must the f(a), g(a) and D values be to ensure that this acceptable situation is true. Unfortunately there is no unique inverse to the function relating f(a), g(a) and D to failure. This means that there are an infinite number of combinations of f(a), g(a) and D to give one Pf. However practical considerations would limit the range to quite a small subgroup. But the important thing to note is how all the functions are interrelated. Therefore if a range of failure probabilities are selected it is possible to set out a practical array of f(a), g(a), and D's that give rise to this range.

5. The Value of ISI

The previous section does nothing for us, as it stands. It merely concludes that a given set of values for the principal parameters leads to a unique failure probability, or, given a failure probability a sensible range of principal parameters can be inferred. Let us consider each of these sets as individual sets and call them 'Probability of Failure Sets (PFS)' thus :-

$$PFS_i = \{(f_i(a),\ g_i(a),\ D_i,\ h_i(a)\}\ Pf_i)$$

Clearly we would like to think that a set giving a low Pf was representative of the real situation. The problem is justifying that set and it is to that end that the ISI is to be aimed.

In section 2 the first question asked was "Given that we believe a certain situation exists, what would we expect to find during an ISI?". Now we have a full set of possible situations, the set of PFS, against which to ask that question. Assume now that an ISI has been carried out and a number of defects 'm' have been observed. The follow up question "How does this knowledge affect the original belief" now has some meaning against all the possible PFS. Such a situation is generally known as a Bayesian condition.

6. Probability of Finding a Defect During ISI

The preceeding sections leave us having to determine the probability of finding 'm' defects during ISI. This requires an inspection efficiency curve, let this be I(a).

The function I(a) used to describe the inspection efficiency will depend on the type of inspection and the area being inspected. It will not depend on any individual PFS. However what is seen, or not seen, by I(a) will depend on these individual sets in that it will be a function of the defect density and distribution entering service, plus the crack growth that has occurred up to the particular ISI being considered. This then provides the conditional probability required in the Bayes analysis.

It is not the objective of this paper to become deeply involved in the question of inspection efficiency, although clearly it is an important part of any ISI analysis. It has however been recognised that the most difficult thing to achieve is an accurate defect sizing. The model has therefore been set up to consider one of the following three situations :-

1) Only the total number of crack like defects is known,with no sizes.
2) All defect sizes are known.
3) A split situation where below some given size only the number of crack like defects is known, but above that given size any defect can be given a specific size.

7. Samples from a Group

Inherent in this method is the assumption that any one sample can be considered as a sample from an identifiable group and is representative of that group. Further that the family group is amenable to a statistical representation.

For example we might consider a particular set of welds which were all made to the same welding procedure and see a similar loading history through life. There may be N welds per plant and P plants giving NP welds. Any one weld chosen from these NP welds will have its own unique number of defects, its own unique crack growth behaviour and will have seen a unique loading history. However all of these welds are manufactured in the same way of the same material and are all subject to a similar loading through life. This implies that we can consider them as one family and that this family is amenable to a statistical representation. Hence any one weld is just the realisation of a random variable which is a unique member of the whole family. Information from that one sample is now applicable to the statistics that describe the family. In this way it is possible to establish a satisfactory description of the whole family with only a limited sample.

The above serves to illustrate the point made in section 2, that this approach to ISI does not try to show any reduction in failure probability as a result of ISI. Instead it tries to use the ISI data to establish what the true Pf is for this particular family.

The preceeding establishes that a given family of components can be treated statistically. What it does not do is tell us whether or not a set of components can be considered as a family. This can only be done by the user and places a strong commitment on him to justify that a given set of components can be considered as a family.

8. The Mathematics of the Problem

The previous sections tried to lay down the basic engineering concepts behind the method. This section carries through the detailed mathematics of the method.

8.1 Definitions

8.1.1 $f(a|\underline{\Theta})$ is the p.d.f. of initial crack like defect sizes 'a' (which lie in a component of thickness 1), which depends on a set of parameters $\underline{\Theta}$ (with $\underline{0}<a<a_m$ and $a_m<1$). The type of p.d.f. that has been considered to date has been Weibull.

8.1.2 The rate at which crack like defects exist at the start of life is δ per component.

8.1.3 The critical crack size (that size above which failure will occur) is x_c.

8.1.4 The function $g(a, t \mid \underline{\alpha})$ gives the size of a crack at plant age t which was a size 'a' at plant age zero, where $\underline{\alpha}$ are the parameters of the function.

8.1.5 I(a) is the inspection efficiency : i.e. for a given defect of size a, I(a) is the probability of detecting that defect if the component in which the defect lies is inspected.

8.1.6 There are N components in each plant and at an inspection n of these are inspected.

8.2 Assumptions

8.2.1 The mode of failure is via continuous crack growth (along the axis perpendicular to the component surface) to size x_c from a defect that was present in the component at time zero (i.e. no crack initiation during the plants life time).

8.2.2 There is no plant effect, i.e. the parameters which describe the mode of failure are the same on each plant.

8.2.3 No re-inspection occurs and the components to be inspected are chosen randomly.

8.3 Prior Knowledge

Let $\underline{V} = (\underline{\Theta}, \underline{\alpha}, \delta)$. The true value of $\underline{V}$ is not well known so a probability density function (pdf) is constructed which should reflect our knowledge of the situation prior to the accumulation of any plant data that will be used in the updating process. The method used to date, has been to try and cover the space of all reasonable values of $\underline{V}$ by a finite number $\underline{V}_1, \ldots, \underline{V}_L$ and assign a prior probability P_i^*

to each $\underline{V}_i$ (where $\sum_{i=1}^{L} P_i^* = 1$).

8.4 Updating the Prior Knowledge

As data is collected our knowledge on the values of $\underline{V}$ are updated i.e. the values of P_i^* are changed.

Data used to update the P_i^* is : no failures have occured up to time "t", a plant is inspected at time t and m cracks are found and, possibly, the sizes of these m cracks.

8.4.1 Updating Via Non-Failure

By "non-failure" in this context we assume that all the cracks in the system are less than the component thickness. This is a pessimistic assumption since it has been assumed that x_c < thickness. The updating procedure is carried out in time steps of $t_k - t_{k-1}$ where t_{k-1} was the time of the previous inspection (of a plant) and t_k is the time of the present inspection. For each plant the times t_{k-1} and t_k are converted into the age of the plant at t_{k-1} (T_{k-1j}) and the age of the plant at t_k (T_{kj}) where j=1 ..., M with a total of M plants.

Let Z_{ijk} be the size of an initial crack on j^{th} plant that would just grow to full component width at time t_k assuming that Vi is the correct set of parameters.

i.e. $1 = g\,(Z_{ijk},\ T_{jk} | \underline{\alpha}_i)$

then the probability of "non failure", $r_{ij}\,(t_{k-1}, t_k)$, on the jth plant between the age T_{k-1j} and T_{kj} given that there was "non failure" to age T_{k-1j}, that $\underline{V}_i$ is true and there is 1 crack in the plant, is given by,

$$r_{ij}(t_{k-1}, t_k) = \int_0^{Z_{ijk}} f(a|\underline{\Phi}_i)\,da \Big/ \int_0^{Z_{ijk-1}} f(a|\underline{\Phi}_i)\,da$$

using the Poisson distribution with rate $N\delta_i$ for the number of cracks ℓ gives the conditional probability $P_{ij}\,(t_{k-1}, t_k)$ of non failure on the jth plant given $\underline{V}i$ is true as

$$P_{ij}\,(t_{k-1}, t_k) = \sum_{\ell=0}^{\infty} r_{ij}\,(t_{k-1}, t_k)^{\ell}\; e^{(-N\delta)}\;(N\delta_i)^{\ell}$$

$$= e^{(-N\delta_i\,(1 - r_{ij}(t_{k-1}, t_k)))}$$

Therefore the probability Q_i of "non-failure" on all the M plants given $\underline{V}_i$ is true is

$$Q_i\,(t_{k-1}, t_k) = \prod_{j=1}^{M} P_{ij}\,(t_{k-1}, t_k)$$

These $Q_i\,(t_{k-1}, t_k)$ are then used to update our knowledge on the V_i using Bayes Theorem. Let $R_i\,(t,w)$ be the probability that $\underline{V}_i$ is correct at time t having previously had w inspections (so $P^*_i = R_i(t,0)$) then

$$R_i(t_k, w) = \frac{R_i(t_{k-1}, w)\,Q_i(t_{k-1}, t_k)}{\sum_{S=1}^{L} R_S(t_{k-1}, w)\;Q_S(t_{k-1}, t_k)}$$

The $R_i\,(t_k, w)$ represents our knowledge of Vi at time t_k after w inspection.

8.4.2 Updating Via Inspection

Let k^{th} inspection (which is on the jth plant) be carried out at time t_k, and the age of the jth plant at t_k be T_{jk}. Then the probability q_i of not finding a crack given that one is there, that V_i is true and that there are no cracks present greater than the component thickness is given by

$$q_i = 1 - \int_0^{Z_{ijk}} f(a|\underline{\Theta}_i)\ I\ (g(a,T_{jk}|\ \underline{\alpha}i))da \ \Big/ \int_0^{Z_{ijk}} f(a|\underline{\Theta}i)da$$

If n components are inspected then the probability of finding r cracks (which are not sized) given that $\underline{V_i}$ is true is given by S_i where

$$S_i = \sum_{u=r}^{\infty} \frac{u!}{r!(u-r)!} (1-q_i)^r\ q_i^{u-r}\ \frac{e^{-n\delta_i} (n\delta_i)^u}{u!}$$

$$S_i = \frac{\left((1-q_i)n\ \delta_i\right)^r e^{-(1-q_i)n\delta_i}}{r!}$$

i.e. Poisson with rate $(1-q_i)n\delta_i$

If the r cracks are sized as $x_1,\ldots.., x_r$ then $a_{i,1},\ldots.. a_{ir}$ are calculated such that

$$x_b = g\ (a_{ib}, T_{jk}|\ \alpha_i) \qquad b = 1,\ldots.., r$$

then S_i is set at the liklihood of finding r cracks $x_1,\ldots..., x_r$ given that $\underline{V_i}$ is true

$$S_i = \frac{e^{-(1-q_i)n\delta_j} (n\delta_i)^r}{r!} \prod_{b=1}^{r} f(a_{ib}|\underline{\Theta_i})I(x_b)$$

Again Bayes theorem is used to update our knowledge on the $\underline{V_i}$. The updated probabilities $R_i(t_k,k)$ after the kth inspection taking place at time t_k on the jth plant is given by

$$R_i(t_k,k) = \frac{R_i\ (t_k,k-1)S_i}{\sum_{u=1}^{L} R_u(t_k,k-1)\ S_u}$$

9. Summary and Conclusions

A statistical method has been proposed to evaluate the benefits of ISI. The method is based on Bayesian logic using the information obtained from ISI on a particular family of components to update a vague prior belief about that family. The vague prior belief consists of separate sets of defect density, defect size distribution, in service crack growth and critical crack size. If no defects are found during ISI, the method builds up confidence that the true probability of failure for that family is very low. If defects are found the method indicates which of the sets is most likely to be correct, and hence what the true distribution of failure probability is for that family of components.

The sets must cover the range, most optimistic, most likely (best estimate) and worst possible situation. A distinct advantage inherent in this, is that it is not necessary to be able to identify what the actual, or true, situation is; only that it lies within the range of the chosen sets. It is the ISI data that is used to try to identify the true situation.

The model can be used retrospectively on inspections carried out to date. It can then be used as an ongoing method of assessing the most likely true situation regarding the probability of failure of any given family of components and thus direct future ISI programmes in the most effective way.

Since the model assumes that a given family of components is amenable to a statistical interpretation, it is necessary only to consider a sample of that family to provide information about the whole. This in turn implies that only a limited or reduced ISI programme is required to provide the necessary information. It should however be clearly understood that if the outcome of a limited ISI programme indicates that the probability of failure is too high, increasing the number of ISIs or the sample size per ISI, will not necessarily change the situation. It will provide more data which will increase the confidence in the failure distribution but this may only serve to confirm that the situation is unacceptable. If this becomes the situation the basic premise of the analysis must be changed to ask, how does ISI affect the probability of failure. It would be possible to develop the model toward this objective.

The model does not implicitly include any form of consequence ranking. This can only be applied to the model by designating different acceptable levels for the failure probability for different family groups. But here it must be realised that the model attempts to determine the true situation and cannot therefore be artificially made to meet a given target, without corrupting the ISI results input to the model. Thus it must be concluded that the question of consequence is external to the model but can be included in the analysis. If the model is to be used to plan and optimise a future ISI programme, assumptions about the outcome of the ISIs must be made. In this context the consequence of failure must be used in order to determine what probability of failure is acceptable to any given family or components. The model can then be used to optimise the ISI programme against a given confidence level for the prescribed failure probabilities. Note that this does not mean that the confidence level will be reached, since this depends on the realisation of the ISI programme. It does however provide a meaningful way of planning.

The method of analysis proposed in this paper is still under development. It is however at a point were it can be used to provide initial estimates of the benefits of ISI and to plan a future ISI programme. The future development will not change the basic principles laid out in this report. The authors believe that the proposed method provides a meaningful way of interpreting the findings of an ISI programme.

RISK-BASED INSPECTION – DEVELOPMENT OF GUIDELINES

*Volume 1
General Document

*Volume 2 – Part 1
Light Water Reactor (LWR) Nuclear Power Plant Components

*Volume 2 – Part 2
Recommended Inspection Program for LWR Nuclear Power Plant Components for ASME Code Section XI Consideration

Volume 3
Fossil Fuel-Fired Electric Generating Station Applications
(In course of preparation)

Volume 4
Petroleum Refinery Processing and Storage Applications
(In course of preparation)

Volume 5
Noncommercial Nuclear Facility Applications
(In course of preparation)

*These documents also to be published by the U.S. Nuclear Regulatory Commission, respectively, as:

NUREG/GR – 0005	Vol. 1
NUREG/GR – 0005	Vol. 2 – Part 1
NUREG/GR – 0005	Vol. 2 – Part 2

LIST OF PUBLICATIONS ASME CENTER FOR RESEARCH AND TECHNOLOGY DEVELOPMENT

Volume No.	Title
CRTD-1	*Consensus of Current Practices for Lay Up of Industrial and Utility Boilers,* 1985, 28 pp., Book Number H00336
CRTD-2	*Research Needs in Thermal Systems,* 1986, 221 pp., Book Number I00212
CRTD-3	*Goals and Priorities for Research in Engineering Design,* 1986, 164 pp., Book Number I00230
CRTD-4	*Pressure Systems Energy Release Protection (Gas Pressurized Systems),* 1986, 425 pp., NASA Contractor Report 178090
CRTD-5	*Thermodynamic Data for Biomass Materials and Waste Components,* 1987, 496 pp., Book Number H00409
CTRD-6	*Analysis of the Feasibility of Replacing Asbestos in Automobile and Truck Brakes,* 1988, 136 pp., Book Number I00264
CRTD-7	*Hazardous Waste Incineration: A Resource Document,* 1988, 192 pp., Book Number I00266
CRTD-8	*Research Needs in Dynamic Systems and Control – Volume 1: Strategic Research Initiatives and Opportunities,* 1988, 58 pp., Book Number I00275
CRTD-9	*Research Needs in Dynamic Systems and Control – Volume 2: Acoustics and Noise Control,* 1988, 48 pp., Book Number I00276
CRTD-10	*Research Needs in Dynamic Systems and Control – Volume 3: Control of Mechanical Systems,* 1988, 40 pp., Book Number I00277
CRTD-11	*Research Needs in Dynamic Systems and Control – Volume 4: Machine Dynamics,* 1988, 37 pp., Book Number I00278
CRTD-12	*Research Needs in Dynamic Systems and Control – Volume 5: Nonlinear Dynamics,* 1988, 32 pp., Book Number I00279
CRTD-13	*Peer Review Thermal Energy Storage Program,* 1989, 46 pp., Available from CRTD, Washington, D.C.
CRTD-14	*The ASME Handbook on Water Technology for Thermal Power Systems,* 1989, 1900 pp., Book Number I00284
CRTD-15	*Research Needs and Technological Opportunities in Mechanical Tolerancing,* 1990, 51 pp., Book Number I00298
CRTD-16	*Engineering Fluid Mechanics Workshop Report,* 1990, 114 pp., Book Number I00299
CRTD-17	*ASME Research Committee on Corrosion and Deposits From Combustion Gases Seminar of Fireside Fouling Problems,* 1990, 183 pp.
CRTD-18	*ASME Ash Fusion Research Project,* 1990, 11 pp., Book Number I00307
CRTD-19	*Achievements in Tribology,* 1990, 173 pp., Book Number H00629
CRTD-20–1	*Risk-Based Inspection – Development of Guidelines: Volume 1, General Document,* 1991, 156 pp., Book Number I00311

www.ingramcontent.com/pod-product-compliance
Lightning Source LLC
LaVergne TN
LVHW082338200926
848481LV00012B/3383